D1636396

PROBLEMS IN MODERN GEOGRAPHY

Derelict Land

PROBLEMS IN MODERN GEOGRAPHY

Series Editor — Richard Lawton *Professor of Geography, University of Liverpool*

PUBLISHED

J. Allan Patmore	*Land and Leisure*
David Herbert	*Urban Geography: A Social Perspective*
Kenneth Warren	*Mineral Resources*
John R. Tarrant	*Agricultural Geography*

IN PREPARATION

Description in Human Geography	David M. Smith
Factorial Ecology	W. K. D. Davies
Marketing Geography	R. L. Davies
Maps in the Computer Age	Ian S. Evans and David C. Rhind
New Communities	H. Brian Rodgers
North Sea Oil and Gas	Keith Chapman
Nuclear Power and the Energy Revolution	P. R. Mounfield
Political Geography of the Oceans	V. R. Prescott
The Rebuilding of Europe: A Geographical Analysis	Mark Blacksell
Recreational Geography	Patrick Lavery (editor)
Water Supply	Judith Rees
World Automotive Industry	G. T. Bloomfield
World Steel	Kenneth Warren

PROBLEMS IN MODERN GEOGRAPHY

KENNETH L. WALLWORK

Derelict Land

ORIGINS AND PROSPECTS OF A LAND-USE PROBLEM

David & Charles Newton Abbot London
North Pomfret (VT) Vancouver

In memoriam
Maureen Wallwork
1934–1972

ISBN 0 7153 6741 2
Library of Congress Catalog Card Number 74–82829

Set in 11 on 13pt Times New Roman and printed in Great Britain
by Latimer Trend & Company Ltd Plymouth
for David & Charles (Holdings) Limited
South Devon House Newton Abbot Devon

Published in the United States of America
by David & Charles Inc North Pomfret Vermont 05053 USA

Published in Canada by Douglas David & Charles Limited
3645 McKechnie Drive West Vancouver BC

Contents

List of Illustrations

PLATES

IN TEXT

List of Tables

Preface

> In a crowded dirty little country like ours one takes defilement almost for granted. Slag heaps and chimneys seem a more normal, probable landscape than grass and trees
>
> George Orwell. *The Road to Wigan Pier* (1937)

> The scars left behind by industrial development of the past, the abandoned waste heaps, disused excavations and derelict installations and buildings no longer needed by industry, are an affront to our concept of an acceptable environment in the 1970s
>
> The Rt Hon Peter Walker, MBE, MP,
> Secretary of State for the Environment (1971)

THIS book is largely about the development of derelict land as an accompaniment to various forms of extractive and mineral-based industries. It also examines the attempts to minimise future dereliction, both by technical developments and by stricter legal controls on mineral exploitation and waste disposal. The work concludes with an examination of progress made in reclaiming derelict land and the prospects for eradicating dereliction. There can be little doubt that prevention of dereliction, or, failing that, immediate restoration of worked sites, is an accepted aim of mineral planning in the western world. But it is equally important to realise that dereliction is not going to be prevented by the simple expedient of ceasing to work minerals, although the choice of alternative locations for deep mines and open pits may increasingly reflect environmental as well as purely economic decisions. (In one sense environmental considerations have a direct and immediate bearing on costs of mineral working, when landscaping of the mine has to be borne as an additional production cost.)

Most of the examples used are drawn from Great Britain, a country with a long history of dereliction and a much shorter, but none the less significant, history of reclamation. Illustrations are also taken from overseas, both to include examples not found within Great Britain, notably extensive opencast working of low-grade ores, and to demonstrate the universality of some forms of dereliction. The emphasis throughout is on dereliction as a feature of the landscape of industrial regions, its modes of occurrence, spatial distribution over time, and its restoration for a variety of beneficial uses. Readers requiring technical explanations will find the bibliographies helpful.

Note on Measurements

Metric measurements are used throughout this book. Where the original figures were approximations, the nearest metric equivalent has been given (eg 'a coal seam about 3ft thick' has been rendered a 'coal seam about 1m thick'). Otherwise an accurate conversion has been employed. Areas are in hectares (ha): 1 hectare equals 2·47 acres. Weights are in metric tons: 1 metric ton equals 0·984 long tons.

PART ONE

DERELICT LAND IN ENGLAND AND WALES: GENERAL SURVEY

CHAPTER ONE

The Definition and Classification of Derelict Land

DEFINITIONS OF DERELICT LAND

IN legal terms derelict land is that which has been abandoned by its owner, a definition which, were it to be used as the sole criterion of dereliction, would drastically reduce the scale of the problem. However, the apparent simplicity of the legal and literal meaning of dereliction is misleading, and derelict land has proved to be more difficult to define than might be supposed by the casual observer of the landscape. The problem of definition stems largely from reconciling an economic with an aesthetic approach. Thus it is possible for land to appear to be spoiled or degraded without its being economically derelict: the waste heap is as much a part of the industrial plant as the factory buildings or the mineral workings which gave rise to it. Conversely, but perhaps less commonly, land which is economically derelict may give no offence to the eye, for often this will comprise non-industrial forms of land use such as derelict woodland.

The need to define dereliction precisely became important in Great Britain once it had been decided to frame legislation designed both to curb its further growth and to provide financial aid for its reclamation. The first official British definition was that coined originally in setting the terms of reference for S. H. Beaver's pioneer survey of the Black Country in 1945, and subsequently used as the basis for other investigations of the problem. Derelict land was defined as

> Land which has been so damaged by extractive or other industrial processes or by any form of urban development that in default of special action it is unlikely to be effectively used again within a reasonable time, and may well be a public nuisance in the meanwhile.

When it was decided to collect annual returns of derelict land from local authorities in 1964, a much shorter official definition was produced: in it derelict land was considered to be 'land so damaged by industrial and other development that it is incapable of beneficial use without further treatment'.

Succinct though this second definition is, it is commonly agreed that the statistics based upon it are inaccurate to the extent of excluding at least half the derelict land in England and Wales. The reasons for this are twofold: firstly, there are errors of classification in collecting the data at local level; secondly, there are several categories of land which are excluded from the returns even though the definition would seem to embrace them. Both points are referred to below (pp 20–24), and although the first might be resolved by collecting data centrally, the second would require a major change in the principles guiding the annual survey. There have, therefore, been attempts to produce definitions of derelict land which might more realistically reflect its true extent, more particularly by introducing aesthetic considerations. J. R. Oxenham (1966) defined derelict land as that 'which has been damaged by extractive or other industrial processes (and) in its existing state is unsightly and incapable of reasonably beneficial use'. P. W.

Bush (1969) observed that 'Derelict land . . . gives offence to the eye, it has an ugly and unkempt appearance'. Definitions of this kind may accord with the popular image of dereliction, but they cannot lead to an objective assessment of its scale as a land-use problem.

That the visual impact of dereliction should be thought to be important is not surprising, for there is a long tradition of literary description which equates industrialisation with squalor and degradation of the landscape. Whether literature shapes public opinion or reflects it makes no difference to the fact that such views exist, and in consequence many of the 'old' industrial areas of Great Britain labour under the disadvantage of a poor environmental image, which is said to discourage new capital investment and to impede economic regeneration. All writers on the industrial scene have not, however, unequivocally related industrial development to visual squalor. Some, notably Arnold Bennett and George Orwell, have employed forms of romantic imagery which, for example, permit comparisons to be made between a smoking colliery waste heap and Stromboli. The two extremes of attitude are brought together by W. G. Hoskins (1955), who observed of places such as St Helens, Lancashire, that 'the landscape of Hell was foreshadowed' in the nineteenth-century industrial scene, but that the abandoned mines of north Cornwall when 'seen towards sunset (were) vastly melancholy and mysterious'.

Clearly any definition of derelict land which invokes aesthetic considerations is likely to produce ambivalence of this kind. This is not to minimise the importance of the visual impact of dereliction and its unfortunate social and economic side effects, but until studies of landscape evaluation and environmental perception have become more objective, it would be unwise to introduce aesthetic concepts as part of definitions of derelict land. At the same time it has to be remembered that the official British definition has presented difficulties of interpretation in spite of the fact that it contains no aesthetic criteria. Criticisms have been made of its failure to include actively used sites which are visually offensive and may in future become derelict – which hardly seems to be fair comment when the purpose of the definition is to identify current dereliction. The ambiguity of the terms 'other developments' and 'beneficial use' has also been criticised, but from a practical point of view it is claimed that a more rigorous definition would restrict rather than enlarge the range of derelict land qualifying for state aid towards reclamation.

More serious criticism can be levelled at inconsistencies in the application of the definition, particularly when it forms the basis for statistical surveys. There are variations in the mode of collection of the raw data by local authorities which may come to light when the returns for successive years are compared but more frequently are undetected. In particular the returns for successive years may show a reduction in the amount of derelict land without indicating that any reclamation has taken place in the intervening period. This may reflect genuine reappraisal and resurvey, but it may also be a product of purely administrative change. Thus derelict land to which planning permission has been attached, or abandoned railway land disposed of to other landowners by the British Rail Board, will normally be excluded from the official returns even though it has not been reclaimed or restored to beneficial use. Although it is claimed that the annual surveys have increased in accuracy since their initiation in 1964, the problems inherent in standardising returns collected locally remain, and more consistent data are not likely to be achieved unless the information is collected centrally, possibly on the basis of regular interpretation of aerial photographs.

There are, however, other reasons which explain the shortcomings of the official statistics, for there are specific exclusions from the returns of land that would otherwise be termed derelict. The six categories of exclusion each refer to land which has been damaged by some form of development. As their existence adversely affects the accuracy of the official returns, it is necessary to examine them in some detail. They are

1 land to which after-treatment conditions apply;
2 land in current use continuing under the General Development Order 1963;
3 land which, although not in current use, is subject to planning permission for future development;
4 infilling sites, war damage, and urban clearance schemes awaiting redevelopment;
5 land which has blended into the landscape in the course of time, or has been put to some acceptable form of use;
6 land derelict from natural causes.

Category 1 applies to most active mineral working sites established after July 1948, for the original permission to extract will normally have contained after-use conditions of some kind. However, it is widely acknowledged that the existence of planning conditions which specify restoration or other remedial works will not necessarily prevent further dereliction from occurring. Legal sanctions may be virtually unenforceable, particularly if the company which originally agreed to the terms has since gone into liquidation. Even if this is not so, it may be claimed that working has not ceased because part of the site is still in use while the remainder is to all intents and purposes derelict. Or, it may be claimed that activity is only temporarily in abeyance, pending either a resumption of work or the initiation of new developments (a contingency also covered by Category 3). Whatever the circumstances, none of the land will be classified as derelict while the planning conditions still apply to it, however unlikely their fulfilment may be.

Category 2 exists because successive General Development Orders have excluded from planning control excavations and waste heaps that were in existence in July 1948, *and continued use of them*. While it would not have been easy to apply after-use conditions to existing sites, the extension of absolution from planning control to developments at these sites has greatly exacerbated the problem. It has proved to be particularly damaging in the coalfields, where it is estimated that about 25 per cent of all colliery waste heaps lie outside planning control. The effect of Category 2 on the statistics is again to exclude entire sites from the classification of derelict land, even though only part may be in active use at the present time.

Category 3 reveals a further weakness in planning control over mineral working, for any site which is covered by planning permission for development in 'the foreseeable future' is, however derelict it may now seem to be, excluded from the statistics. This exclusion covers sites which may be temporarily unworked and those for which some alternative form of development is proposed. There is no way in which such development can be enforced, so that land may lie idle for lengthy periods of time without being added to the official total of derelict sites. It is possible that the placing of time limits on planning permissions may end this anomaly, for the Department of the Environment assumes that such sites will be included in the returns once the planning permission has lapsed.

Taken together the first three categories of exclusion probably account for the more serious discrepancies between the official and unofficial returns; well supported estimates suggest that there is at least twice as much dereliction as the official statistics reveal. Thus a survey carried out by the West Riding of Yorkshire County Council in 1966 revealed 6,850ha of derelict land, compared with an official return of 2,540ha. Most of the discrepancy fell into the classes 'excavations and pits' and 'other forms of dereliction', where the categories of exclusion were most likely to be represented. There is an obvious legal difficulty in terming derelict the land excluded in the first three categories, for the existence of planning controls, or the fact that sites are in partial or intermittent use, places them outside the scope of state-aided reclamation projects, and this is the main purpose of official recognition. On the other hand this is land which would be termed derelict in the absence of any legal qualification, and its inclusion in the statistics would be more realistic than, say, an extension of definition to incorporate subjective aesthetic considerations.

Category 4 is perhaps the exclusion which presents fewest distortions to the official record, although the term 'infilling' may be susceptible to a wide range of interpretations. Much of the land in this category comprises major urban clearance sites which, however derelict they may appear to be for varying periods of time, are almost certain to be built over or put to other authorised uses within the immediate future. The cleared sites of abandoned factories awaiting future planned development fall into the same category. It is true that some localities may bear the imprint of 'planning blight' well in advance of clearance and rebuilding, and there are instances in which uncertainty about the implementation of a particular programme of building will have given rise to a protracted period of this form of semi-dereliction. However, problems of this kind are normally short lived, and even the exceptional cases of lengthy 'planning blight' do not produce difficulties comparable with those experienced in areas of serious industrial dereliction.

Category 5 comes nearest to including aesthetic judgements in determining the extent of derelict land, for ideas may vary as to when dereliction has 'blended into the landscape', and 'can reasonably be accepted as part of the natural [sic] surroundings'. Land so excluded will frequently have been abandoned by industry many years ago, when the amplitude of relief on derelict sites was slight, even though the area covered may have been large. In most instances there will be little or nothing to reveal the former existence of industrial activity. Similarly this category also extends to sites which, although not reclaimed in any way, have been put to an acceptable use, such as fishing in or sailing on wet mineral excavations. Again aesthetic considerations may be invoked, for the setting of such features varies appreciably and in many cases it is the quality of the surrounding landscape which helps to determine the degree to which an 'acceptable' use has been established. It is arguable that some of the land falling into *Category 5* may be capable of more economic use, however inoffensive it may now appear to be, and to this extent the full extent of derelict land in a given locality may be masked by exclusions under it.

Category 6 poses interesting if somewhat marginal problems of definition, akin to those involved in determining the difference between an Act of God and human negligence. Natural dereliction, widely defined, might include all land which had ceased to be agriculturally productive, ranging from abandoned uplands to unkempt estuarine marshes, and including for convenience derelict woodland. Although such a definition would be useful in an attempt to measure the loss of land to agriculture by abandonment over a period of time, its breadth precludes its use in any survey of derelict land as commonly understood. However, there are forms of 'natural' dereliction which are directly attributable to industrial activity, mainly permanent flooding, or formation of marshland as a result of mining subsidence or other forms of interference with drainage systems stemming from waste tipping and excavations. Damage caused to vegetation by atmospheric pollution is not properly speaking an aspect of 'natural' dereliction, although it may be found in conjunction with derelict land.

The identification of damage caused by mining subsidence is not always easy, particularly outside built-up areas, where its very existence may be difficult to detect. The landforms produced by mining subsidence do not differ greatly from those which result from natural slumping. Formation of marshland and permanent flooding may result from bad drainage management rather than damage by undermining, and although it may not be difficult to establish the facts at the present time (bearing in mind the point that mineral operators may not lightly accept responsibility for such damage where positive proof is lacking), it is not easy to assess the extent of this kind of dereliction when attempting to reconstruct earlier phases of land use (see pp 31–2). There is also some confusion as to whether land flooded after subsidence is classifiable as derelict. The Civic Trust (1970) categorically states that it is not, but the North West Economic Planning Council (1969) equally firmly notes its inclusion in the returns, and the Department of the Environment states that such land is to be recorded in the category 'other forms of dereliction'. It is clear that the existence of Category 6 specifically excludes 'natural dereliction', but this does not resolve the problem of deciding what is natural in this context.

It is evident that the principal weakness of the official British definition of derelict land lies not in its failure to recognise the visual aspect but in its list of exclusions. Although these are almost entirely justifiable in terms either of planning procedures or of policy on the award of grants in aid of reclamation, the exclusions are the main source of persistent inaccuracy in the official returns. Although the Department of the Environment proposes to change the mode of collecting data in England, in order to remove some of the anomalies referred to above, the fact remains that the official statistics employed in this account relate only to the hard core of derelict land. Any attempt to examine the total amount of derelict land produced by industrial and related forms of economic activity requires the adoption of a somewhat wider definition. The threefold division outlined below follows the official definition in all but two important respects: firstly, the various exclusions relating to planning conditions and consents are ignored, along with those relating to some forms of natural dereliction; and secondly, a category termed potential dereliction is added, comprising the more obvious sources of future derelict land. The sub-divisions of derelict land are as follows:

1 *extant dereliction:* land now derelict, as officially defined, but including sites covered by planning conditions and permissions;
2 *potential dereliction:* land that will, on cessation of present use, become derelict unless remedial action is taken;
3 *partial dereliction:* (a) regressive – land which has been derelict and, although never reclaimed, has been put to other uses; (b) progressive – land which has begun to deteriorate and will become derelict unless remedial action is taken.

Each of these sub-types of derelict land can be classified and further elaborated as a basis for examining the development of a derelict landscape through successive stages of change. *Extant dereliction* needs no further explanation. *Potential dereliction* is, perhaps, less explicit, for the inclusion of a site in this category does not mean that it is bound to become derelict, although it would do so were no remedial action to be taken. On the other hand some forms of land use which are excluded from the definition might become derelict on the basis of past experience. This applies to abandoned industrial sites unconnected with mineral working, and transport installations. The category *potential dereliction* includes the most obvious sources of future derelict land, such as colliery waste heaps, and quarries and pits of various kinds, but it would be unrealistic, for example, to include elements of the railway and canal system, unless they were solely used by forms of activity that were themselves classified as potentially derelict. *Partial dereliction* represents the two extremes of the problem. On the one hand it comprises land in the initial stages of degradation, notably that damaged by mining subsidence. On the other it includes sites which have been disturbed by industrial activity but have been put to fresh uses without any form of remedial action. These uses may be acceptable but are not necessarily those that a reclamation scheme could have produced. The regressive variant of partial dereliction poses relatively few problems while so much derelict land survives in an unacceptable form, but it is nevertheless an element in the total pattern of dereliction and ought to be recorded.

The term derelict land as employed in this book approximates very closely to the spirit of the official definition (p 18) but without the various categories of exclusion (pp 21–4) that reduce its breadth of interpretation. This means that some forms of land use which may appear to be derelict are excluded, notably poor quality farmland caused by bad management, felled or partly cleared woodland, and transient dereliction related to urban clearance schemes. A more contentious exclusion also has to be made, in that abandoned sites controlled by the Ministry of Defence are omitted partly because of the uncertainties which surround their future use. It is apparent that even by extending the range of the official definition, and without recourse to aesthetic consideration, there is still scope for debate as to what constitutes derelict land. The key considerations must always be that damage has resulted from industrial and related forms of economic activity, or their cessation, and that without remedial action the land so damaged cannot be beneficially used.

SOURCES OF DATA

The problems of definition discussed in the previous section are intertwined with those of establishing facts about the development of dereliction and classifying its modes of occurrence. It is, therefore, necessary to review the principal sources of data, both historical and contemporary, and the classification of land use, before considering the location and extent of derelict land in Great Britain (Chapter 2) and undertaking the case studies which form Part Two of this book.

British land-use statistics in general pose problems of interpretation which have been thoroughly examined, notably by R. H. Best and J. T. Coppock (1962). R. H. Best (1968a, 1968b) has also investigated the transfer of land from agricultural to urban uses, and S. H. Beaver (1968) has paid special attention to changes in industrial use of land. Most of the published estimates of urban and industrial land use cover the whole of England and Wales, or at best relate to the standard regions. Information for smaller unit areas can be deduced from the annual agricultural returns, but with a large margin of potential error stemming from the mode of collection of the data.

The amount of agricultural land recorded can be deducted from the total area of land in a given locality, but the resultant balance is not necessarily urban or industrial land. Apart from obvious non-agricultural and non-urban features such as forests and inland water, the balance may also include the area covered by farmsteads and roads, together with the sum of various forms of inaccurate and ambiguous measurements in the individual farm returns.

Even on a national scale estimates of land use from this source leave 516,000ha (2·3 per cent of the total land surface) unaccounted for, including space occupied by opencast mineral workings. Although the annual agricultural statistics have been available since 1866, no comparable data exist for urban and industrial land uses, save those detailing changes from agricultural to other forms of use, which have been available for England and Wales since 1927, and for Scotland since 1960. However, the weaknesses of these statistics relating to agricultural and urban land use are as nothing compared with the virtual absence of any information on derelict land before 1948.

The first estimate of the extent of derelict land made by J. R. Oxenham (1948) indicated the need for a more complete survey of dereliction, which was carried out by the Ministry of Housing and Local Government in 1954. It had been hoped that further information could be derived from local authority reviews of development plans, but the piecemeal nature of the surveys and reports produced by the planning authorities necessitated a change of policy. From 1964 onwards, each local authority has had to make annual returns of derelict land, proposed reclamation, and reclamation carried out. These data are more fully discussed below (pp 34–6); they provide a better basis for analysis of derelict land than any source previously available, and they are also superior to the forms of information most readily available on other facets of urban and industrial land use over Great Britain as a whole.

Before 1954 no official statistics of dereliction existed other than those collected for relatively small areas by the Ministry of Town and Country Planning, including S. H. Beaver's pioneer survey of the Black Country in 1945, and subsequent surveys of North Staffordshire and the East Shropshire coalfield. Otherwise the principal official references to derelict land appear in reports relating to specific problems, notably those on mining subsidence (1928–9, 1947) and reclamation of ironstone workings (1939, 1946). Important though they are as sources of information, these reports do not shed much light on general problems of dereliction and can only provide a partial picture of a relatively narrow range of problems. It is at first sight surprising that the inter-war years, during which so much dereliction rapidly appeared as the coal and iron industries contracted, yield little official record of the problems posed by derelict land. The reports of the Commissioners for the Special Areas made passing reference to it in the 1930s, but in a way which perhaps explains the general dearth of information, for in 1936 the Commissioner for England and Wales observed that the clearance of derelict land would not be encouraged if no economic advantage stemmed from it. In this context economic advantage meant immediate reoccupation of a site by industry or another source of revenue, so that any incentive to tidy up land merely as a means of providing short-term employment was discouraged. Under these circumstances it is hardly surprising that no body of statistical information about the extent of dereliction at this time exists.

In the absence of land-use statistics it is necessary to turn to other sources of information on the growth of dereliction. The task of reconstructing the landscapes of derelict areas is methodologically similar to any other which endeavours to recreate the past, with one important difference. Dereliction consumes its antecedent forms as it develops, leaving behind few, if any, relict features. While it is true that some localities may not have progressed beyond an early stage of dereliction, this does not mean to say, for example, that a long-abandoned mining landscape in West Durham provides a model of what the Lancashire coalfield looked like before it was transformed by subsequent forms of damage. Thus a present-day area of dereliction will give little or no direct evidence of the earlier stages of its development, whereas a nineteenth-century industrial town will still retain much to inform us of its past, notwithstanding the scale and vigour of recent clearance schemes.

Other differences also conspire to make studies of the history of dereliction more difficult than many other reconstructions and explanations of the recent past. Most notable are the deficiencies in archival material, particularly that relating to the twentieth century, when the rate of dereliction became much more rapid than hitherto. Leases of mining royalties may mention such matters as the provision of support for specific buildings, bridges and lines of communication; and less commonly clauses requiring the restoration of land after working may be included. However, such documents generally contain little of relevance to a study of dereliction, and the records of colliery companies and other mineral extractors equally shed little light on the growth of derelict land.

It was the usual practice of colliery companies to buy much of the land above their workings in order to avoid possible claims against damage from subsidence: land so acquired could also be used for tipping waste without further trace in legal documents. The concern of colliery companies about subsidence was almost entirely directed to problems of support underground and not to its consequences at the surface. Waste disposal was similarly viewed from the standpoint of production techniques, particularly the problem of moving dirt quickly out of the workings between successive shifts. Dereliction scarcely enters the record because it was not considered to be of concern to the mining industry, and although records of value to studies of the continuing creation of derelict land will have stemmed from recent legislation, they are not readily available, nor are there equivalent documents relating to earlier periods. Other extractive industries yield equally little information on dereliction, for their concern was again largely with those elements of the problem which directly affected production. Thus the nineteenth-century Le Blanc alkali manufacturers sought a method of tackling the problem posed by the noxious waste *galligu*, produced at a rate of 2 tons for every ton of finished alkali, not because of its offensiveness and space-consuming propensities but because large quantities of valuable sulphur were locked up in the waste heaps.

Company records are, therefore, generally unrewarding as a source of information on derelict land. Even where oblique references to the problem appear, they are fragmented. However, one source of data provides both national cover and information which can be translated into simple statistics of dereliction – the topographical maps and cadastral plans of the Ordnance Survey. Maps at a scale of 1:10,560 provide adequate detail within limits imposed by the fact that the land-use symbols employed were not designed solely to depict derelict features. Major features such as waste heaps and quarries are easily identifiable, but features such as scrub-covered hummocky ground or marshland require more thorough interpretation. Supplementary evidence, such as earlier editions of the topographical sheet, or geological maps, may be required where the record is not clear. The most difficult features to distinguish are waste heaps in upland localities, where the symbol used is similar to that for scree, and marshland produced by mining subsidence, where the symbol used depicts both 'natural' marshes and those produced as a result of bad management of artificial drainage. Although the maps and plans provide a systematic national cover, the dates of survey within a given edition may vary appreciably, even over relatively small areas. This is particularly true of the 1:10,560 maps and large-scale plans produced on a county sheet basis, but even on the more recent National Grid publications discrepancies in the date of survey may occur within a single sheet. Thus piecemeal revision of the plans from which 1:10,560 maps are produced often means that the map relates to a variety of dates of survey, which may be emphasised by the use of different conventional symbols according to the pattern of revision. Notwithstanding such inconsistencies the Ordnance Survey's maps and plans provide a relatively reliable historical record spanning a century or more, and they are invaluable as a means of tracing the development of the landscape, as well as giving a basis for measuring the area of derelict land at various times in the past.

Topographical maps provide only partial information on land use, but for two periods – the early 1930s and the early 1960s – they are supplemented by land utilisation maps. The published sheets of the first Land Utilisation Survey, at a scale of 1:63,360, did not separately distinguish derelict land, although they can be used to assist interpretation of the 1:10,560 topographical sheets. Unfortunately the field maps of the survey no longer survive, but some supplementary material appears in the county reports. The maps of the second survey at a scale of 1:25,000 (of which 110 sheets had been published by 1971) indicate derelict land, but include urban clearance sites in this category while excluding most 'natural' dereliction. Two other sets of maps are also of value in the study of derelict land, although neither provides the full cover given by the topographical series. The maps of the Geological Survey, particularly those at a scale of 1:10,560 of the coalfields, are invaluable to the interpretation of the origins of derelict land where the record is otherwise obscure. In addition the memoirs and the special publications of the Geological Survey relating to mineral resources provide a great variety of supporting information. Secondly, some of the maps and related memoirs of the Soil Survey separately identify disturbed ground in areas of opencast mining and mineral waste disposal.

A final source of record material comprises photographs of various kinds. Ground-level photographs of mineral workings and the related industrial scene provide evidence of the pre-derelict state of the landscape, and views of the progress of dereliction can be used to clothe the bare bones of cartographic reconstruction with descriptive flesh. Particularly good sets of record photographs survive to illustrate the salt-working subsidences which afflicted mid-Cheshire between the 1880s and the 1920s, and the development of non-ferrous ore mining in Cornwall. Oblique aerial photographs, which first became widely available in the early 1920s, fulfil a similar function, and cover a sufficient time span to illustrate the development of derelict land in certain localities (see, for example, plates, pp 68 and 85). Vertical aerial photographs, available from the mid-1940s, were soon used to identify derelict land without recourse to field survey (Beaver 1969). High quality aerial photographs, produced specifically to assist in the survey of derelict land reclamation schemes largely date from the 1960s, and their use in plotting the extent of current dereliction will be discussed subsequently. Taken together, maps, photographs, and other documentary material of the kind outlined above yield adequate data for the study of the development of derelict land. While the statistics of land use that can be derived from reconstructions of earlier derelict landscapes have neither the range nor the accuracy that would have resulted from contemporary field surveys, they provide a worthwhile substitute for such data.

The study of derelict land at the present time obviously presents fewer practical problems than does historical reconstruction, for sites can be visited and recorded in detail, aerial photography is more readily available and land-use statistics provide a framework for further enquiry. There remain, however, problems of interpretation and classification, and, for the practical planner, decision as to which form of reclamation ought to be adopted in particular instances.

The aerial photograph has replaced the topographical map and the field-by-field survey of land use as the primary research tool. The development of remote sensing techniques in particular is hastening the speed with which basic land-use information can be assembled, and the possibilities of computerised mapping will ensure equally rapid processing of material. It will, however, still be necessary to carry out field checks both to monitor these advanced techniques of recording data and to investigate problems of particular interest at first hand. The use of conventional vertical aerial photographs requires the construction of a photographic key in order to assist the process of interpretation, and this must be checked against field evidence. The study of industrial land use from aerial photographs requires a considerable knowledge of the morphology of industrial plants, and their ancillary services, which can only be gained by a combination of checking obscure points of evidence in the field and using the range of supplementary material already used to elucidate similar problems in reconstructing past landscapes. Thus, although the acquisition of primary data is now much simpler than in the past, the methods of interpretation and explanation remain much the same.

In addition to the greater wealth of material from which a land-use record can be derived there is also the direct information on derelict land contained in the annual surveys carried out between 1964 and 1971 in England, and between 1964 and 1969 in Wales. The survey identifies three simple types of dereliction: spoil heaps, excavations and pits, and other forms of dereliction. For each category an estimate is made of the proportion justifying restoration, and data are also collected recording the amount of land in each category restored during the preceding year and scheduled for restoration in the forthcoming year. The data are published* for each administrative county and county borough, but it is also possible to consult the returns for lesser administrative units at the Department of the Environment, and several of the maps which illustrate subsequent chapters are based on this source. The returns do not identify the types of industry or other activity responsible for particular forms of dereliction and it must be remembered that they are based on the official definition (p 18) and the list of specific exclusions (pp 20–24).

In 1971–2 the Department of the Environment investigated ways in which the scope of the survey might be improved without reducing its value as a means of identifying hard-core dereliction which was likely to attract grants in aid of restoration. After taking advice from interested parties and conducting field trials the Department has introduced a new form of survey to take effect on 31 March 1974. (It was deferred from 1973 in order to take advantage of the reorganisation of local government.) The new survey is based on the same definition as the

* For England they are available in mimeographed form from the Department of the Environment. The statistics for Wales, published by the Welsh Office, are in a slightly different form from that described above. Returns are not collected elsewhere in the United Kingdom, but a survey was also to be introduced in Scotland during 1973.

old, but is more elaborate in four main respects. Firstly the three categories of dereliction have been increased to five by separating 'military and other service dereliction' and 'abandoned British Rail land' from the 'other forms of dereliction' category. The return of land restored since the completion of the previous survey, and the statement of intended restoration during the following year are both subdivided in this way, and for each category information is sought on the source of finance. In addition the survey asks for details of recorded increases in dereliction between successive surveys, partly to distinguish between genuine growth and that stemming from resurvey and reappraisal. In 1974 this question is particularly relevant to the inclusion of service land, previously ignored by the annual surveys, but in future years it will provide a means of monitoring the accuracy of the returns.

Secondly the new survey incorporates questions on derelict land that was previously excluded from the returns because restoration conditions applied to it. Sites covered by planning permissions or other arrangements which 'nominally provide for after treatment but which have been found to be unenforceable or ineffectual' are recorded separately. Thirdly the new survey collects information on the extent of current mineral working, distinguishing sites covered by restoration orders from those not so covered, and identifying thirteen types of mineral (see Table 1). Finally data on sites used for waste tipping (other than mineral waste, which is included elsewhere) are collected, distinguishing public refuse tips from industrial and commercial tips and noting whether restoration conditions apply to the site.

The new survey overcomes many of the weaknesses of the old without reducing its effectiveness in identifying the major current problems of dereliction. The first section of the new survey is in almost every respect identical with the old, using the same definition and list of exclusions, but employing a fivefold subdivision of the returns. However, the inclusion in sections two and three of data on active mineral working and

TABLE 1 Amount of land covered by permissions for surface mineral working and mineral waste tipping where work is in progress or has not yet commenced at survey date

Mineral	*Total area affected by*					
	Soil heaps and tips	*Excavations and pits*	*Plant/buildings/settlement lagoons as appropriate*	*Area restored† since last survey*	*Area affected since last survey*	*Area not yet worked*
Chalk						
China clay						
Clay and shale*						
Coal						
Gypsum/anhydrite						
Igneous rock						
Ironstone						
Limestone						
Sand and gravel						
Sandstone						
Silica and moulding sand						
Vein minerals‡						
Other minerals						

* Including fireclay, ball clay, and potters clay
† Table 5b only
‡ Including tin, copper, lead, silver, zinc, haematite, barytes, calcspar, fluorspar, witherite

Table 1 is based on Table 5 of the Department of the Environment's questionnaire for the 1974 survey of derelict land; Table 5a comprises sites not covered by restoration conditions, Table 5b comprises sites covered by restoration conditions

waste tipping yields much valuable information on the scale and distribution of potential dereliction, as does the clarification relating to planning controls designed to enforce the restoration of derelict land. A pilot survey of this type carried out in Denbighshire in 1971 showed that potential dereliction amounted to 684ha, compared with extant dereliction of 616ha (Jacobs 1972).

METHODS OF CLASSIFICATION

The definition of derelict land determines the range of land uses comprising the total extent of dereliction, but the more detailed subdivision of this population is largely conditioned by the purpose of a particular survey. A comprehensive survey of derelict land would ideally seek information at several levels, as outlined below and illustrated diagrammatically in Fig 1:

1 source of dereliction: eg surface mineral working, deep mining, mineral processing, abandonment of railway;
2 topography of dereliction: eg excavation, subsidence hollow, waste heap, embankment;
3 amplitude of dereliction: the relative relief of derelict sites, the area covered by individual elements;
4 composition of the surface and vegetative cover, including waterlogged conditions;
5 potential for reclamation, in relation to available techniques, costs, and end uses.

(It would be possible to add a sixth level if an objective method of measuring the visual ill-qualities of derelict land were to be developed.)

No existing survey has employed all these elements, though each has figured in one or more recent surveys (Collins and Bush 1969). Topography and source of dereliction have assumed prime importance in most surveys, whereas potential for reclamation has received least attention. Similarly, amplitude of dereliction has been less commonly considered, although it was a key element in S. H. Beaver's survey of the Black Country (1946) and subsequent work on surface mineral working (1955, 1961). Composition of the surface and vegetation cover have also received less attention, yet both are important pointers to reclamation potential, particularly in relation to soil cover and toxicity of waste. Reclamation potential is not, however, solely a matter of the application of engineering techniques to different forms of derelict terrain, it is also a function of economic demand and prevailing social policies. Ideally a price might be put on the cost of alternative forms of reclamation, coupled with an estimate of the return on capital invested, but without the use of advanced techniques of social accounting this could endanger the prospects of reclamation schemes which did not yield immediate and tangible financial benefits. Similarly the view that 'beauty has an economic value and ugliness an economic cost' (Civic Trust, 1964) is not easily demonstrated on a balance sheet, yet without some appraisal of the cost of

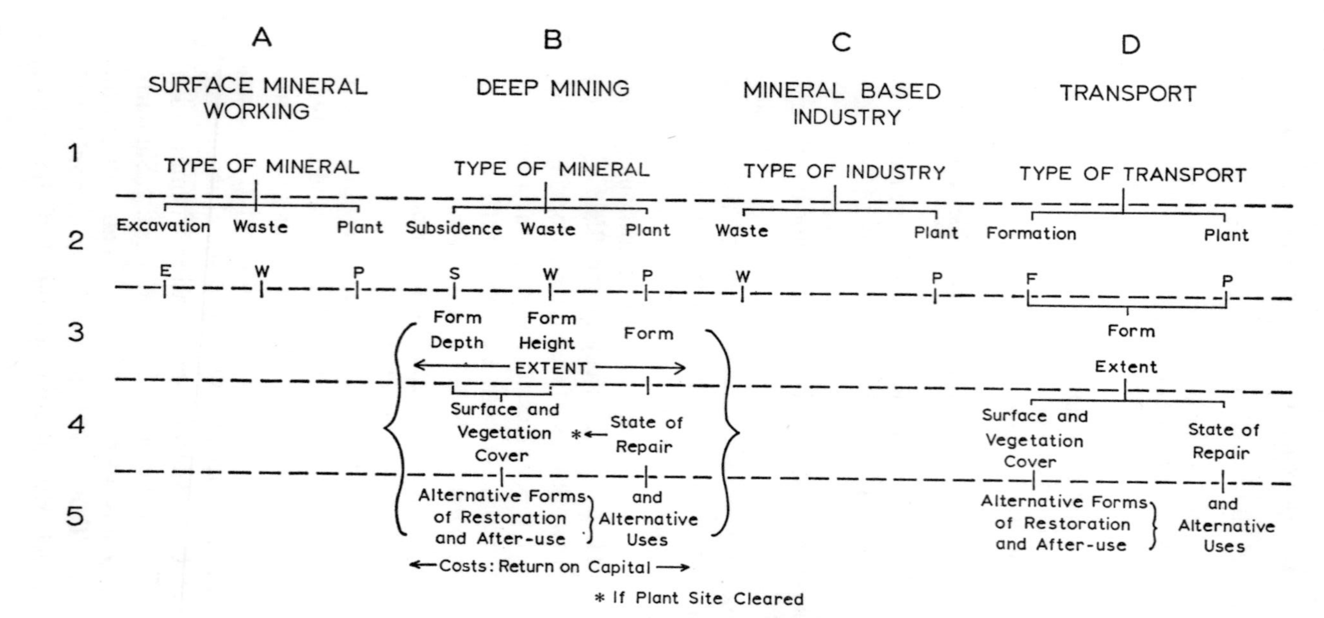

Fig 1 Structure of the classification of derelict land

dereliction based on a scale of disfigurement measured in visual terms the study of derelict land lacks an important dimension. However, no sophisticated technique of evaluating landscape quality whereby this state of affairs might be remedied yet exists (Penning-Rowsell and Hardy 1973).

In studies of the development of derelict land, information on potential for reclamation is less significant than that covered by the higher levels of the classification (Fig 1). The threefold division into extant, potential, and partial dereliction (pp 25–6) can be superimposed on this classification, in that each of the three subdivisions may comprise some common elements. Reconstructions of earlier derelict landscapes largely preclude any detailed measurement of amplitude, surface composition, and vegetation, unless vertical aerial photographs of sufficient quality are available.

Examples of various approaches to the recording of derelict land, using a scheme of notation based on Fig 1 are shown in Fig 2, which comprises a recent survey based on field investigation (Fig 2a), an interpretation based on aerial photography (Fig 2b) and a reconstruction based on topographical maps and cadastral plans (Fig 2c) (see pp 40–1).

References to this chapter begin on p 309.

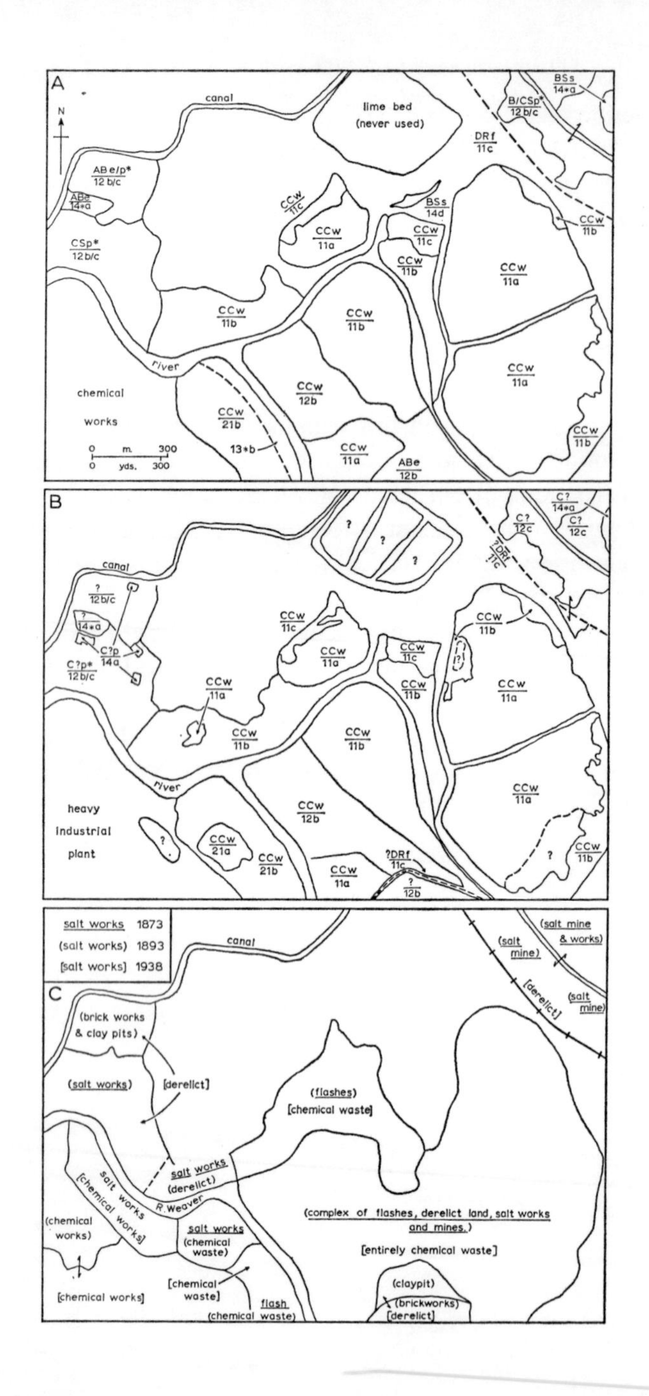
A
canal
lime bed
(never used)
N
DRf
11c
BSs
14*a
B/CSp*
12 b/c
ABe/p*
12 b/c
ABe
14*a
CSp*
12 b/c
CCw
11c
CCw
11a
BSs
14d
CCw
11c
CCw
11b
CCw
11a
CCw
11b
CCw
11b
CCw
11b
river
CCw
11a
chemical
works
CCw
21b
CCw
12b
13*b
CCw
11a
ABe
12b
CCw
11b
0 m. 300
0 yds. 300
B
canal
?
?
?
?DRf
11c
C?
14*a
C?
12c
C?
12c
?
12 b/c
?
14*a
C?p
14a
C?p*
12 b/c
CCw
11c
CCw
11a
CCw
11c
CCw
11b
CCw
11a
CCw
11b
CCw
11a
CCw
11b
CCw
11b
river
heavy
industrial
plant
CCw
21a
CCw
21b
CCw
12b
CCw
11a
?DRf
11c
?
12b
CCw
11a
CCw
11b
salt works 1873
(salt works) 1893
[salt works] 1938
C
canal
(salt mine)
(salt mine & works)
[derelict]
(salt mine)
(brick works & clay pits)
(salt works)
[derelict]
(flashes)
[chemical waste]
salt works
(derelict)
R. Weaver
salt works
[chemical works]
(chemical works)
salt works
(chemical waste)
(complex of flashes, derelict land, salt works and mines.)
[entirely chemical waste]
[chemical works]
[chemical waste]
flash
(chemical waste)
(claypit)
(brickworks)
[derelict]

Fig 2 Derelict land near Northwich, Cheshire, in 1967: (a) from a field survey undertaken by the writer; (b) from an aerial photograph at a scale of 1:7,500; (c) from Ordnance Survey 1:2,500 plans, incorporating historical evidence.

Explanation and Commentary. Figures 2a and 2b employ a scheme of notation outlined in Fig 1 (but omitting level 5). The first capital letter indicates the major category, the second the type of activity (B, brickworks; C, chemical works; S, saltworks; R, railway); the lower-case letter indicates the form of dereliction, as given in Fig 1 (p* indicates site of plant, with few or no surviving structures). The digits relate to components of levels 3 and 4 respectively (all are not used in Fig 2):

Height/depth (above/below surrounding levels)	*Surface configuration*	*Vegetation*
1 under 5m	1 level/gently sloping*	a absent no soil/soil*
2 6m–30m	2 pitted/undulating*	b rough grassland
3 over 30m	3 gullied/steeply sloping*	c scrub/sparse woodland*
4 —	4 waterlogged/flooded*	d reeds

A question mark before notation, entirely doubtful; within notation, absent component doubtful.

The differences between 2a and 2b are marginal, and reflect the following problems: (i) the chemical waste shades from white to a red-brown easily distinguished in the field, but not always clear on the photographs; (ii) minor details identifying sites of industries seen at ground level are not distinguishable on the photographs; (iii) part of the area mapped is inaccessible, but can be more clearly interpreted from the photographs.

Fig 2c summarises the land-use history of the area on the basis of evidence from 1:2,500 plans dated 1873, 1893 and 1938 (provisional revision). For a more detailed map see Fig 37; part of the aerial photograph appears as the plate on p 157

CHAPTER TWO

The Origins of Derelict Land

ALTHOUGH derelict land may stem from a variety of causes, the most important relate to mineral extraction and the early stages of processing minerals. In Part Two of this book specific examples of derelict landscapes and their mode of origin are examined in detail, but it is also necessary to look at the pattern of development in more general terms. Two related topics are examined in the greater part of this chapter: the scale of British mineral production during the past century or so, and the development of mineral working and processing, in terms of both technical progress and locational change related to the creation of derelict land. In addition there is a brief review of other forms of dereliction.

MINERAL EXTRACTION AND PROCESSING

Although the forms of mining and the relative significance of various minerals may change over a period of time, modern industrial nations cannot survive without the products of mineral working. The most drastic way of preventing dereliction – complete cessation of mining and quarrying – is therefore the least likely to be adopted as a means of solving the problem. However, dereliction is not an inevitable accompaniment to mineral extraction: the opposite view was still readily countenanced as recently as the 1930s but it has become progressively less acceptable. Thus, although dereliction continues, there is now greater interest in prevention and reclamation than ever before and hopes are entertained that the problem of derelict land may have been solved in Great Britain by the early 1980s.

While statistics of mineral production provide a very approximate guide to the changing pattern and volume of dereliction over the years, there is no constant relation between output of minerals and the amount of derelict land resulting from extractive and refining processes, either over a period of time or for a given product worked at different sites at a particular moment in time. Dereliction results from the interaction of a complex set of influences, notably the geological structure and physical location of mineral outcrops, techniques of extraction and refining, and methods of waste disposal related both to technical competence and to the economics of alternative modes of operation. Thus the amount of potential dereliction may rise in relation to the amount of overburden moved rather than to tonnage of mineral quarried, or new techniques of screening and refining minerals may produce an increase in the volume of waste residues that is independent of changes in recorded output of the finished product. There are also problems posed by the statistics themselves. Regular returns of production from establishments covered by the inspectorate of mines and quarries date from 1873, but exclude the output of workings less than 6m deep. Changes in the classification of minerals have been made from time to time, and there are minor inconsistencies stemming from the fact that returns for the whole of Ireland were included in the statistics before 1921. Notwithstanding these various problems of interpretation the statistics of mineral production give an indication of the likely source of dereliction and some measure of its changing nature.

Figs 3 to 9 illustrate various aspects of mineral production

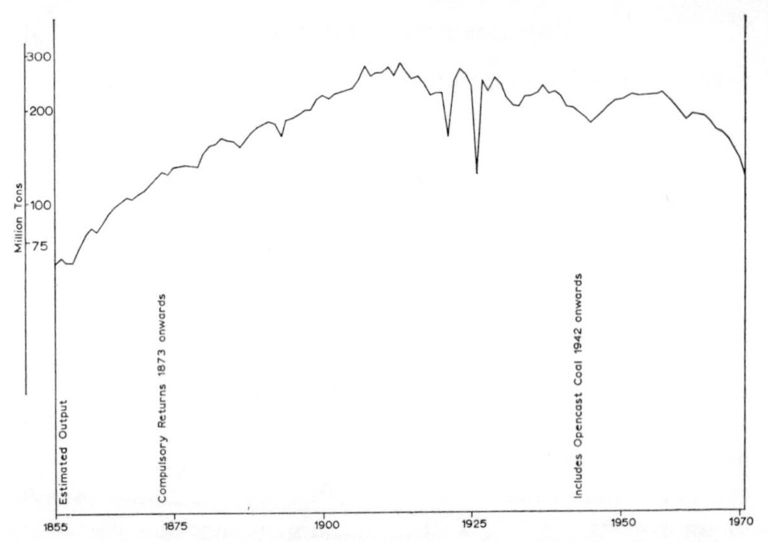

Fig 3 Coal production in the United Kingdom, 1855–1971. Data from various sets of official returns: in Figs 3 and 6–9 outputs have been plotted against a logarithmic scale in each instance

Fig 4 Coal production in the UK by major fields, 1873–1971. Data from various official returns have been aggregated to match as nearly as possible the major divisions employed by the NCB since 1946

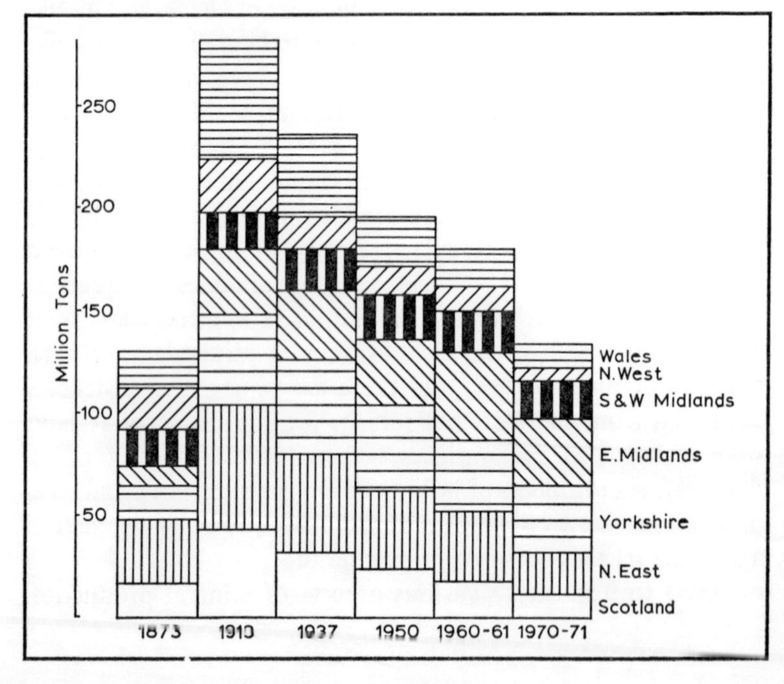

in the United Kingdom during the past 100 years or so. Coal production reached its peak of 292 million tons in 1913, and throughout much of the period between 1919 and 1939 oscillated about the 225 million tons per annum mark. Since 1945 the production of deep-mined coal has generally fallen below this level, and in 1970–71 was 135 million tons. These national totals mask important inter- and intra-regional variations in output which have a bearing on the spread of dereliction, and are discussed below. In spite of the contraction of coal mining the industry still makes a sizeable contribution to the creation of potentially derelict land, and provides an excellent example of the fact that technical advances in mineral extraction may yield increasing amounts of waste at a time when saleable output is falling. In addition the sharp contraction of coal mining since 1958 has produced much derelict land, including abandoned waste heaps, pit-head buildings, transport installations and, in many instances, ancillary plant such as washeries, coke ovens and brickworks. On the other hand the continuation of opencast coal mining, introduced as a short-term emergency measure in 1942, has produced no long-term dereliction, for it has always been the policy to reclaim such sites after use.

Iron ore mining and quarrying have not suffered the same decline as coal mining, for output in 1970 (12,297,000 tons) was little different from that of 1919 (Fig 5). The areas of working have, however, changed far more dramatically than those of coal mining. In 1919 only 50 per cent of British-mined iron ore came from the Jurassic deposits of the Midlands; 20 years later the proportion had risen to 80 per cent, and by 1970 98 per cent came from this region. The shift in location was paralleled by a switch from deep mining to opencast working, but though this initially posed great problems of dereliction, these have been resolved since 1951 (see pp 242–3).

If coal and iron can be regarded as the major forms of nineteenth-century mineral working, the principal raw materials of the construction industry may be considered to be their twentieth-century equivalents in terms of volume of output.

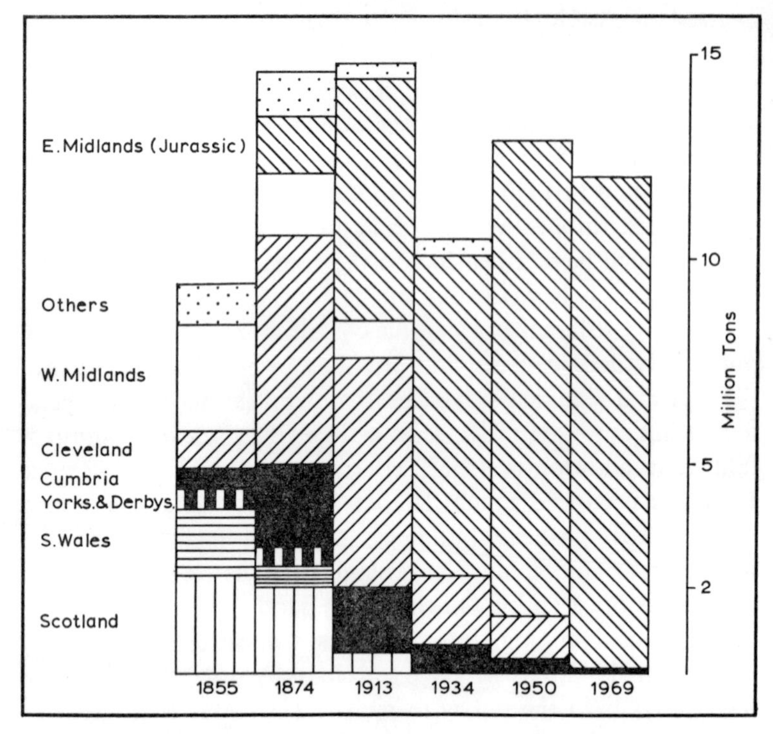

Fig 5 Iron ore production in the UK by major areas, 1855–1970. The decline of the coalfield iron-mining districts (particularly in Scotland, South Wales and Staffordshire) is noteworthy, as is the rise of the Jurassic orefield in the East Midlands

With the exception of slate, all have greatly increased in volume of production since 1919, more particularly the quarrying of sand, gravel and limestone (Figs 6 and 9). Finally several less common minerals have made significant but highly localised contributions to the problem of dereliction (Fig 7): china clay working still creates massive waste heaps and deep quarries, particularly in southern Cornwall, whereas oil-shale mining no longer causes damage, although the cessation of working in 1962 produced an immediate increase in the amount of derelict land in West Lothian. Salt is worked in two forms: rock salt, production of which has increased dramatically since 1945, and

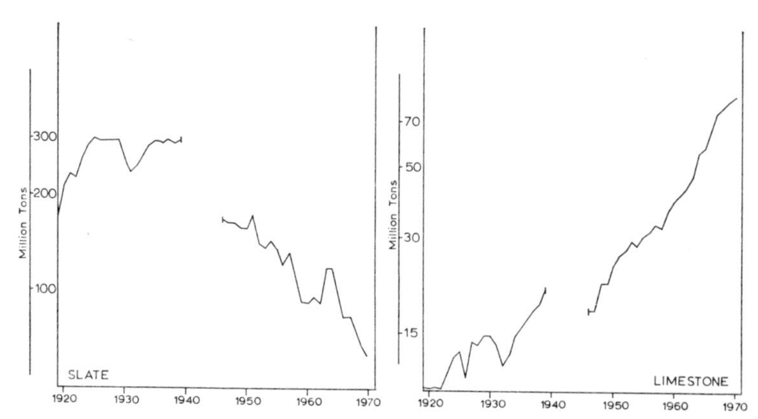

Fig 6 Slate and limestone production in the UK, 1919–70. The decline of the slate-quarrying industry has added to the volume of dereliction in many areas of outstanding scenic beauty. Similarly the major areas of increased limestone production are in scenically attractive localities. (In Figs 6–9 production during the period 1940–45 has been omitted)

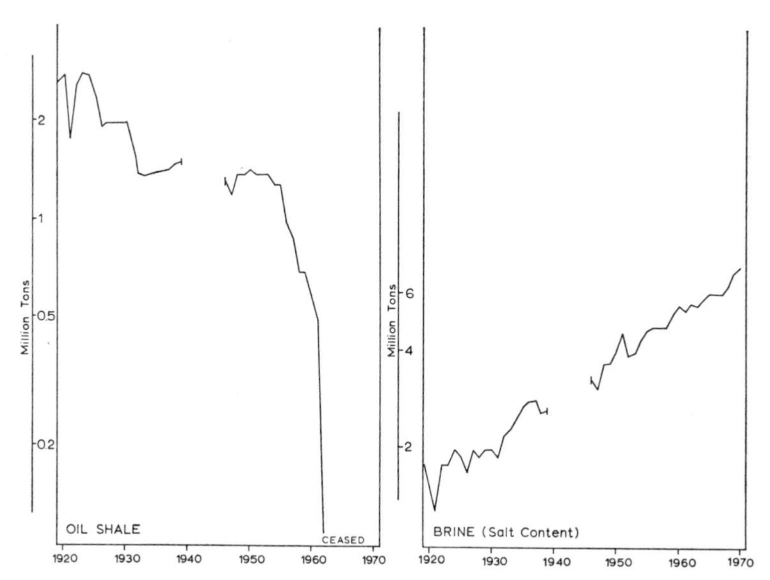

Fig 7 Oil-shale and salt production in the UK, 1919–70. The mining of oil shale, in central Scotland, also yielded large quantities of waste: although the industry ceased production in 1962 the problems left by the massive waste heaps still remain. On the other hand the equally highly localised production of brine (roughly 80 per cent comes from the Cheshire saltfield), which has progressively increased since the 1920s, is not now attended by major dereliction, owing to the introduction of more sophisticated methods of abstraction from the 1920s onwards

brine, which has been pumped at an increasing rate during the same period. Although both forms of extraction formerly caused much dereliction, the problem has abated with the introduction of new techniques of working, an important instance in which a greater volume of mineral production has not been accompanied by an increase in the amount of derelict land.

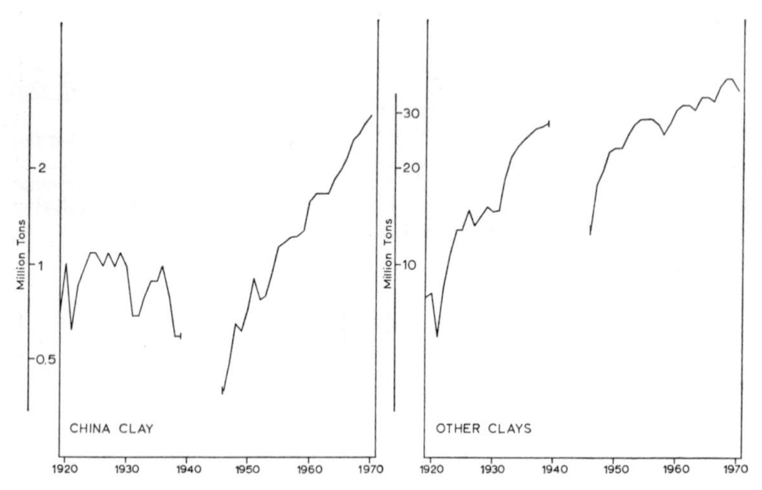

Fig 8 Clay production in the UK, 1919–70. The steady post-war increase in china-clay production, highly localised within SW England, has had important consequences in creating extensive potential dereliction

In addition to mineral working, several sources of mineral-based dereliction result from the processing of ores and other minerals. Iron ore smelting and the refining of various non-ferrous metal ores were formerly sources of much potential dereliction through the need to dispose of waste residues. For a variety of reasons the problem scarcely exists now, and many of the waste heaps of previous years have been removed because they now have an economic value, mainly as road metal or other filling material. Similarly, but for very different reasons, those branches of heavy chemicals manufacture based on mineral raw materials are no longer a major source of potential dereliction.

Page 49 (*above*) Parkhouse colliery, North Staffordshire, 1965, showing conical spoil heap and earlier flat wagon-tips; (*below*) Derelict colliery spoil heaps, Pilsley, Derbyshire. This site forms part of the reclaimed area depicted in plate on p 86

Page 50 (*above*) Jurassic ironstone quarries near Corby, Northamptonshire, 1970. The stages of working are clearly identifiable: the overburden is removed to permit extraction of the ore; and behind the advancing quarry face 'hill and dale' is formed by dumping the waste, which is then reclaimed; (*below*) China clay pits and spoil heaps between Carbis Common and Stenalees, St Austell, viewed from the north-west, 1950. Note the flooded workings and the gullied tips, some of which have a thin vegetational cover

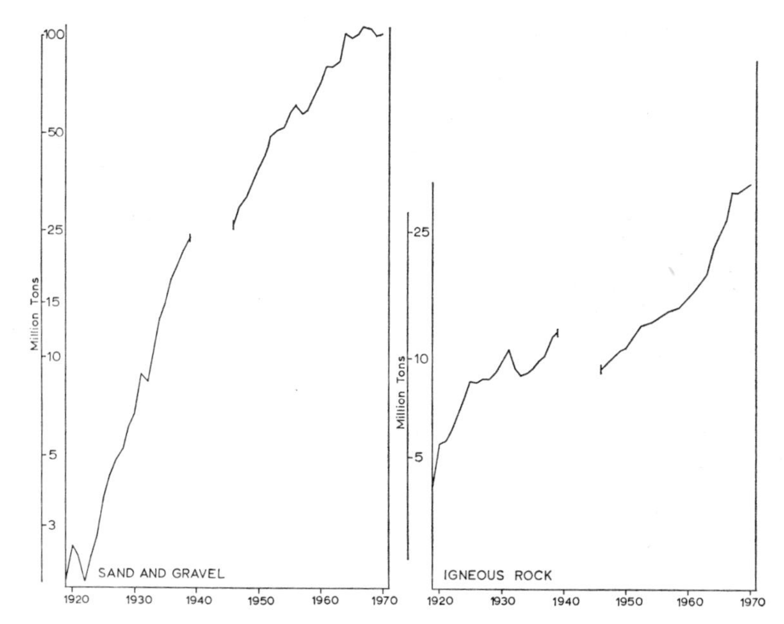

Fig 9 Sand and gravel and igneous rock production in the UK, 1919–70. The almost uninterrupted growth of sand and gravel production since 1919 has posed one of the major problems of potential dereliction, often in localities that were unscarred by nineteenth-century mineral working. The growth of production at igneous rock quarries presents problems analagous to those in limestone-working areas

DEVELOPMENT OF DERELICT LAND IN MINERAL WORKING AREAS

The development of derelict land can conveniently be divided into the three main periods outlined below in Table 2. Each period reflects major changes in the techniques of mineral working and the demand for minerals, and each broadly distinguishes a phase of differing intensity of dereliction. However, with so many industries involved, and a great variety of locations covered, it must be borne in mind that while these phases of dereliction have a general relevance, particular cases may not exactly fit into this framework.

Table 2 Historical Periods of Dereliction

1 *Pre–1800.* Small-scale mineral working, low incidence of dereliction, few major long-term problems of derelict land.
2 *1800–1920.* (i) 1800–70: gradual increase in depth and size of mines; abandonment of some early mining areas; growth of mineral-using industries as a source of dereliction.
(ii) 1871–1920: increasing use of machinery in mining and quarrying; greater depth of mining with advent of new techniques; parallel growth of greater damage from subsidence and waste disposal. Abandonment of many non-ferrous metal mines; first closures of large ironworks and ancillary plants.
3 *1921–71.* (i) 1921–47: increase in extant dereliction due to further closures of mines, quarries, and related industrial plants. Continued growth of potential dereliction in traditional mining areas, and also in new localities of opencast working.
(ii) 1948-onwards: greater importance of legislation designed to minimise dereliction and secure reclamation of land. Accelerated closure of mineral workings, but continued development of the larger, better placed units, with a consequent rise in both extant and potential dereliction.

1 Pre-1800

If the current definition of derelict land were to be applied to sites disrupted by the earliest periods of mineral extraction and refining, very few instances of long-term damage requiring special action would come to light. The flooded peat diggings of Norfolk (the Broads), the slag heaps of medieval iron works, and the numerous overgrown quarries from which cathedrals, castles and colleges drew their stone would not appear in an inventory of derelict land. Yet in their day, and particularly in relation to contemporary technology, all were derelict on abandonment, and have only become acceptable, often inconspicuous features of the landscape through the passage of time. Most pre-nineteenth-century mineral workings were, however, on so limited a scale that they were unlikely ever to have posed more than a highly localised short-term land-use problem. Before 1800 derelict land in the modern sense scarcely existed, but there were instances, particularly during the eighteenth century, in which the bases of future problems were established. Thus the accumulation of slate waste from the mines at Blaenau Ffestiniog began in 1755, and from the Penrhyn Quarries in 1780 when a single large enterprise replaced the earlier small workings. The first rock-salt mining subsidence occurred in mid-Cheshire in 1750, and by 1790 brine-pumping subsidence was also being recorded. Along the north crop of the South Wales coalfield opencast coal and clayband iron ore workings had produced much disruption by 1800, to which was added the growing slag heaps of the ironworks, the first of which was established in 1756. Further south the copper smelting mills of the lower Swansea Valley had developed strongly since their foundation in 1717, and across the Bristol Channel the Cornish landscape was being transformed by the mining of ores to supply these and other works. For the most part, however, there was little intractably derelict land in 1800, and even where potentially severe dereliction was in prospect its scale was still relatively modest.

2(i) 1800–70

During the period 1800–70 several important developments in mineral extraction and processing extended the range of localities that were ultimately to experience problems of dereliction, without at this stage greatly augmenting the area of extant derelict land. In some instances pockets of derelict land with low amplitude of relief were created, notably by the abandonment of primitive mines and shallow quarries which had exhausted the most readily available mineral outcrops. Such localities were characterised by a high turnover among the closely packed small workings, so that dereliction was interspersed with active mining, as, for example, in parts of the Black Country and the non-ferrous metal mining districts of Cornwall. Where mining had penetrated new localities, or was able to tap reserves at greater depth in established coalfields, the workings were larger and more widely spaced than hitherto, and they produced greater quantities of waste for surface disposal. But for as long as most mining was manual, employing the pillar and stall system, and while waste tipping was likewise not fully mechanised, potential dereliction was still not severe.

Quarrying generally presented even fewer difficulties than mining during this period, for the scale of operation was often small, limited as much by the restricted market area before the advent of the railways as by technical considerations. Consequently there were numerous shallow pits and quarries supplying materials for local consumption. For example, many towns and villages had brickfields on their outskirts, using a great variety of deposits to supply the local demand for bricks. Although little is known about the majority of these workings, it is clear that their scale of development posed few long-term land-use problems, and many of them were subsequently built over as the towns which had called them into being expanded. The opening up of the Jurassic ironstone fields of the Midlands in the 1850s introduced a new element of quarrying, but again workings were shallow, and as they were rapidly restored to agriculture, their disruptive effect was minimal. The major large quarries were those working slate in Wales, which had grown rapidly from the 1840s as railway development had greatly enlarged their market area. Even so, the major result of this expansion was the creation of potentially derelict land rather than extant dereliction during the years before 1870.

The expansion of mineral-based industries in the first half of the nineteenth century stimulated mining and quarrying, and the industries caused dereliction themselves. Although there were many instances in which mineral working and mineral processing occurred close together, a major feature of the mineral-based industries was the growing extent to which they were capable of influencing developments in mining some distance away. Thus ironworks might draw their ore from more than one locality in Great Britain and, from the 1830s onward, copper smelters were not only refining British ores but were also importing ore from Cuba and Chile. Blast furnaces and copper smelting mills produced large quantities of slag which had to be dumped on the surface – a man-made lava which, unlike its natural counterpart, did not weather into a fertile soil. Perhaps the most prolific source of potential dereliction was the alkali industry, using the Le Blanc process introduced to Great Britain in 1802. Not only did a Le Blanc works consume about 850 tons of coal, limestone, iron pyrites and salt for every 100 tons of finished alkali made, it also yielded 200 tons of noxious waste. Although an extreme case, the Le Blanc alkali industry demonstrates the effect of mineral users, as well as mineral producers, in the creation of potentially derelict land.

The general location of potentially derelict land in the 1840s (a little later in Northumberland and Durham) is shown in Fig. 10, which is based largely on an analysis of the 1 : 63,360 maps of the period. While it is impossible to measure the detailed extent of derelict land accurately from such a source, the map indicates the major areas with mining, quarrying and mineral using industries that were likely to give rise to dereliction. The importance of the more accessible parts of the exposed coalfields

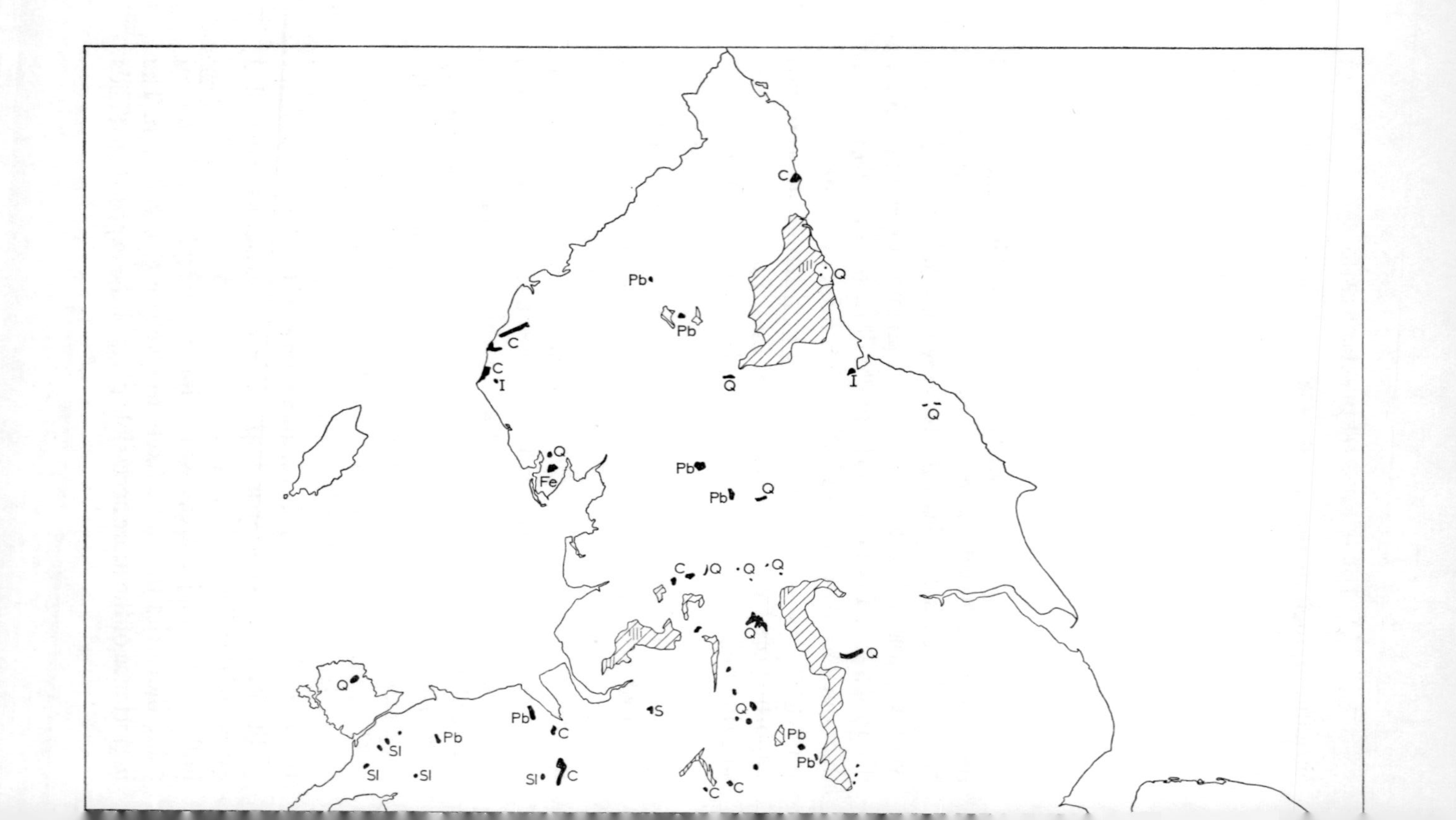
C
Q
Pb
Pb
C
C
I
Q
I
Q
Q
Fe
Pb
Pb
Q
C
Q
Q
Q
Q
Q
Q
S
Pb
C
Pb
Sl
Sl
Sl
C
Sl
Q
Pb
Pb
C
C

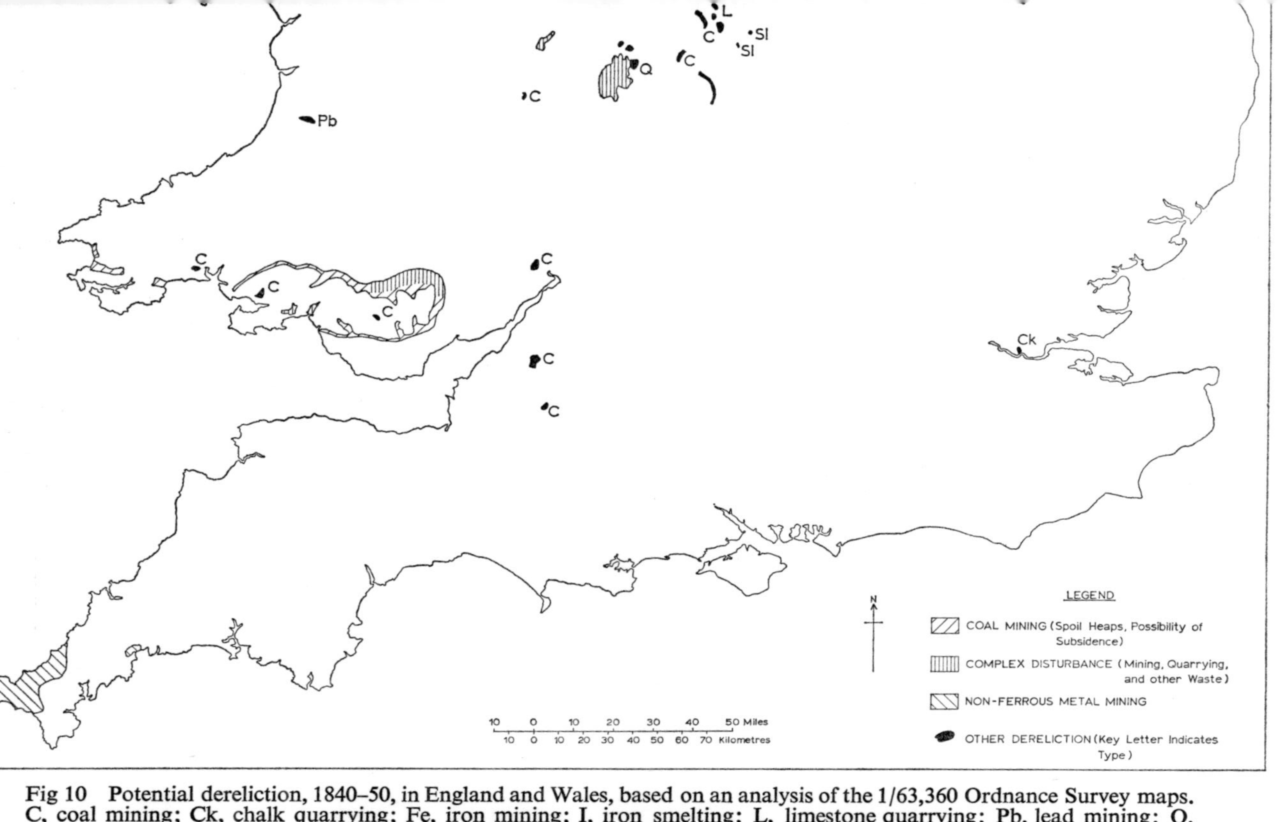

Fig 10 Potential dereliction, 1840–50, in England and Wales, based on an analysis of the 1/63,360 Ordnance Survey maps. C, coal mining; Ck, chalk quarrying; Fe, iron mining; I, iron smelting; L, limestone quarrying; Pb, lead mining; Q, quarrying (general); S, salt working; Sl, slate mining and quarrying

is noteworthy, for these localities not only supported coal mining but were also major areas of iron ore mining and smelting, brick, tile and pottery manufacture, and the production of heavy chemicals. Outside the coalfields the major mining areas were those working non-ferrous metal ores, notably in the Pennines, mid and north Wales, and particularly in Cornwall. Quarrying rarely appeared to be a major source of potential dereliction at this time, except for the Welsh slate workings and the quarrying of limestone. The latter was, however, often related to adjacent developments in iron smelting or chemical manufacture, forming an outliner of coalfield complexes of mines, quarries and mineral-based industries.

2(*ii*) *1871–1920*

This period may conveniently be regarded as one of transition between the primitive stages of development of extractive industries and their highly mechanised present-day state. Although no single date stands out as crucial, several events during these years helped to set the pattern for the least tractable elements of dereliction. The year 1920 is taken as the terminal date because it marks both the end of the post-war boom that was seen as a reversion to pre-war standards of industrial expansion, and the beginning of a major phase of economic difficulty which had important repercussions in the creation of derelict land.

In the coalfields the successful introduction of mechanised mining, the increased adoption of the longwall technique of extraction, and improvements to coal screening and washing all contributed to a growing problem of waste disposal (see pp 116–25). As mechanised forms of tipping spoil were also introduced, this proved not to be an operational difficulty, but it did mean that the waste heaps were often created in forms that made them less amenable to reclamation than had hitherto been the case. Collieries were commonly being sunk to depths of 500–600m, which had previously been exceptional, and were larger and more widely spaced, as were their spoil heaps. In addition coal-mining subsidence had become a more serious problem, largely as a result of the adoption of new faceworking techniques. Consequently the amount of potential dereliction created at this time was both considerable and likely to prove intractable.

Not all mining areas expanded. In the coalfields the areas of decay which had begun to appear before 1870 were more extensive by 1920, notably in the Black Country, East Shropshire and along the northern and eastern rims of the South Wales coalfield. Even in localities still actively developing, some of the earlier collieries which could not be deepened had been abandoned, producing a scatter of derelict land. Where the Coal Measures ironstones had been worked, particularly from the clayband series, there was commonly a trail of derelict pits marking the line of the outcrop. But almost all the extant dereliction in the coalfields was of low amplitude, typified by the 'hills and hollows' of the Black Country, and it was a far less serious long-term problem than the damage being caused in the actively developing mining areas. Other mining districts also contained more derelict land in 1920 than they had 20 years previously, notably through the almost complete cessation of non-ferrous ore mining. Although much of this was also low amplitude dereliction, the waste heaps of tin and copper mines also contained substantial toxic residues. Subsidence was a problem in the majority of mining areas, and included some spectacular collapses of shallow workings in thick deposits, notably in the haematite field of Furness and the Cheshire salt field. The latter area was also plagued by various forms of brine-pumping subsidence which were at their most active in the period 1873–1935.

The quarrying industry still presented fewer problems than deep mining in 1920. Mechanical excavators had been introduced to the Northamptonshire ironstone field in about 1895, but their use was limited until the increased demand for home-produced ore during World War I stimulated their wider adoption. Even so, manually worked iron ore quarries still survived in the 1920s. Brick making was still a widespread industry tied to a large number of small workings, and although the rich resources of the Oxford Clay vale had been tapped since the 1890s by mechanical means, mechanisation of the clay pits was still in its formative stage by 1920. Similarly the china clay workings of Cornwall and Devon were still worked by primitive methods, although the twin problem of deep excavations and high waste heaps was already beginning to assume its distinctive shape. Most other branches of the non-metalliferous mineral working industry had yet to introduce large-scale mechanisation, and the amount of potential dereliction was accordingly small and its amplitude moderate. A major exception was provided by the slate quarries which had, without recourse to elaborate machinery, produced massive workings and, equally important, large mounds of waste material. Alone among the quarries producing materials for the building industry the slate workings had passed their production peak before 1914 (Fig 6), so that the likelihood of abandonment of the least profitable quarries and a consequent growth of extant dereliction was an immediate danger.

Several of the mineral-using industries had become less serious sources of continued potential dereliction since 1870. Non-ferrous metal smelting had increasingly given way to the refining of imported concentrates, and the Le Blanc alkali industry had virtually ceased to exist by 1920. The ammonia-soda process which had largely replaced it as a means of making alkali also yielded waste residues, but at a far less prolific rate than the Le Blanc method. On the other hand ironworks were still dumping large quantities of slag, particularly at some of the newer plants which used lean phosphoric ores. In addition increasing quantities of iron ore were imported for smelting from 1870 onwards, and by the early years of the twentieth century as much pig iron was being produced from foreign as from home ore. There was, however, an indication that the disposal of slag might not always be an unfortunate by-product of iron and steel manufacture. In 1887 it was discovered that slag from basic steel furnaces could be used as a fertiliser. As acid steel production was dominant in Great Britain until 1919, this innovation had little direct effect in the period under discussion, but it was a portent of future improvements. Of greater immediate significance was the fact that technical changes of this kind were helping to cause shifts in the location of iron and steel manufacture. Not only was potential dereliction thereby spread to new areas but the rate of creation of extant derelict land also increased as obsolete plants were abandoned.

The general pattern of distribution of potentially derelict land of various types at the turn of the century is shown in Fig 11, which is based on an analysis of contemporary topographical maps. The coalfields were predictably areas with heavy concentrations of potential dereliction, coupled in some instances with sizeable tracts of low amplitude extant derelict land. The extent of the area susceptible to dereliction had greatly increased since the 1840s (Fig 10) as the mining spread to the deeper seams of the coalfields. In addition the iron ore fields of Furness-West Cumberland and Cleveland had been opened up during the second half of the century. Further new sources of potential dereliction were to be found in localities as widely separated as the china clay workings of Cornwall, the oil-shale and clay workings of Flintshire, and the chalk quarries related to the cement industry of Thames-side. In other long-established mineral-working districts a mixture of extant and potential dereliction was to be found, notably the non-ferrous ore mining areas of Cornwall and the Pennines, the slate quarrying localities in Wales and the Lake District, and the Cheshire salt mining and brine pumping districts.

3(i) 1921–47

The years 1921–47 were characterised by rapid increases in the rate at which extant dereliction was created, coupled with a sustained growth of potentially derelict land. The increase in extant dereliction was very largely a product of the contraction of coal mining, iron smelting, and related forms of industry located mainly, but not exclusively, in what were to be identified

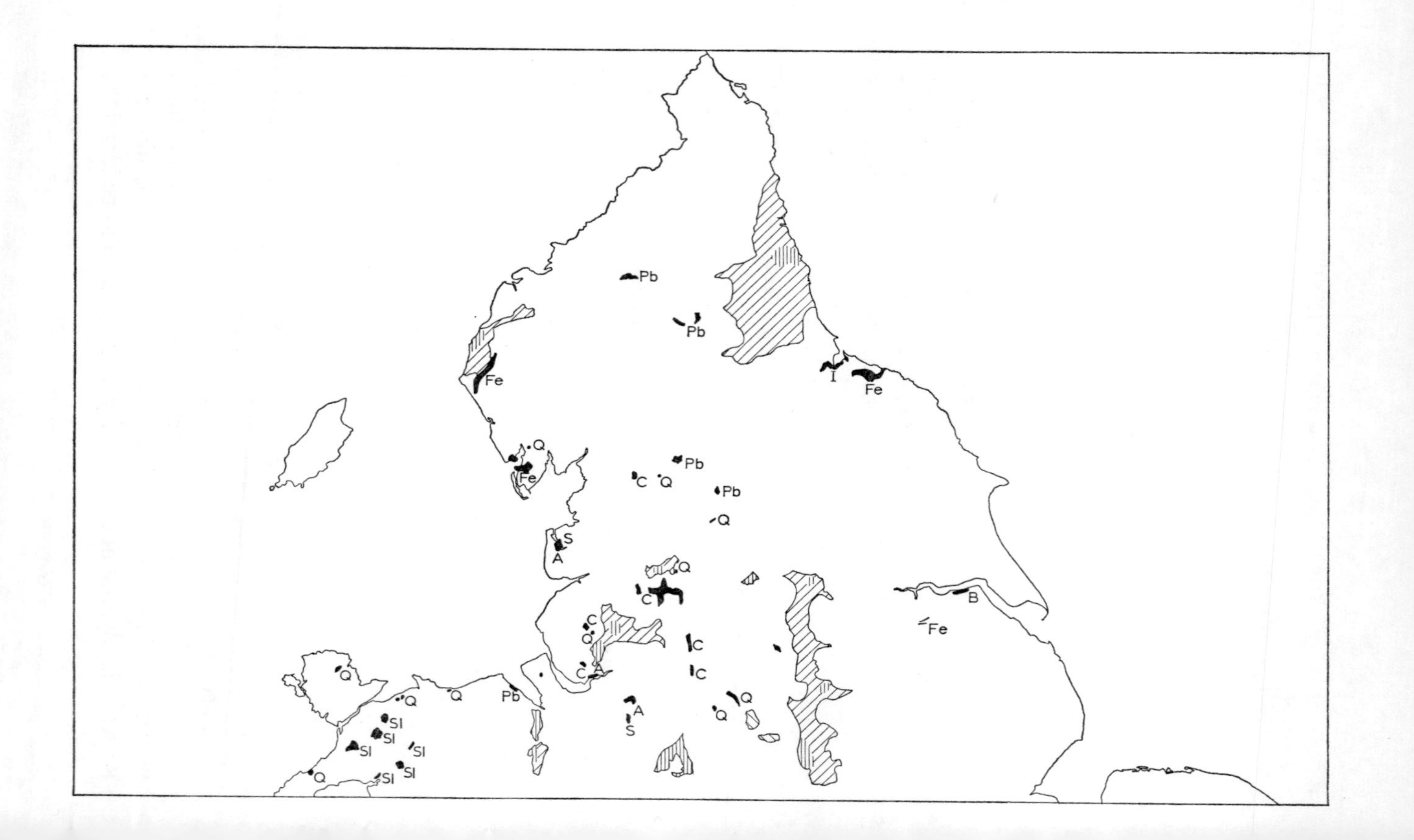

Fe
I
Fe
B
Pb
Pb
Pb
Pb
C
Q
Q
Q
C
C
C
Q
Q
Q
C
Q
C
S
A
A
S
Fe
Q
Fe
Pb
Q
Q
Sl
Sl
Sl
Sl
Sl
Sl
Q
Q

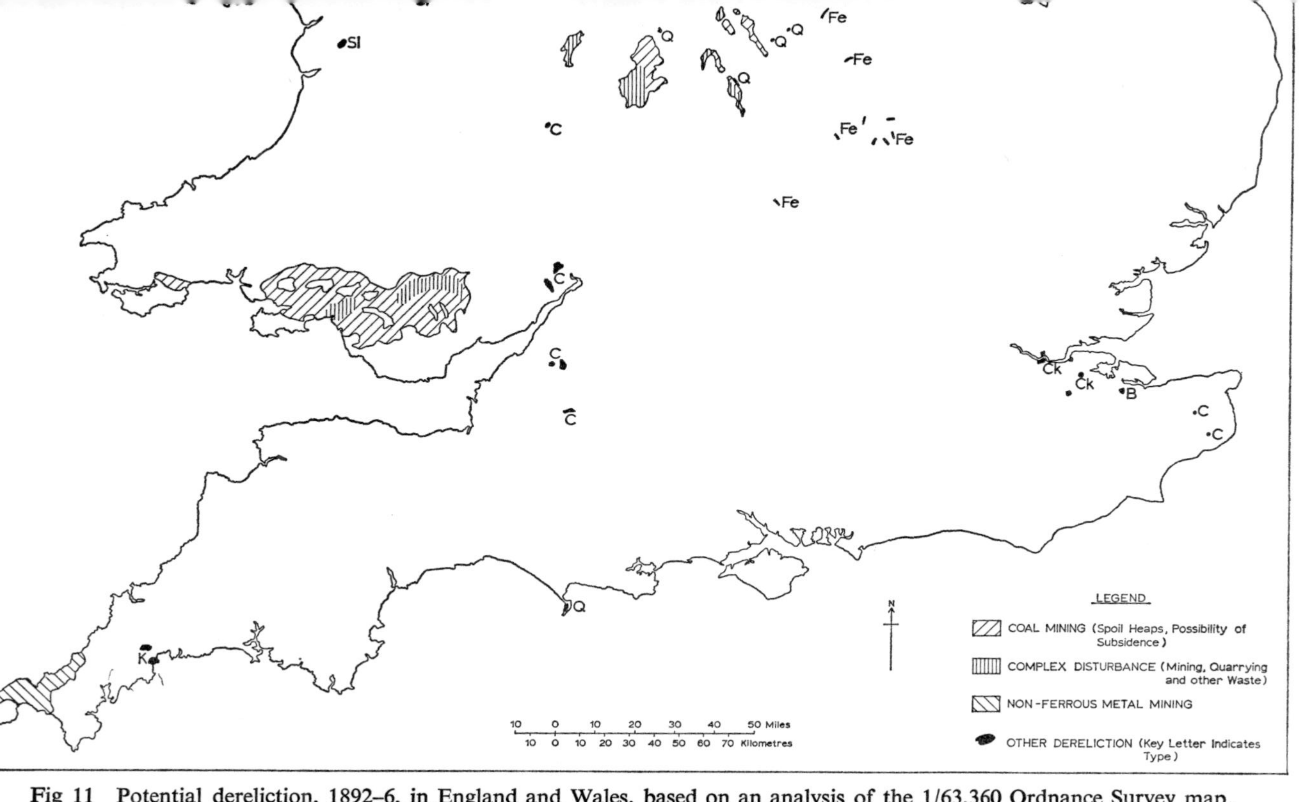

Fig 11 Potential dereliction, 1892–6, in England and Wales, based on an analysis of the 1/63,360 Ordnance Survey map. Key as for Fig 10. A, alkali waste; B, brick-clay working; K, china-clay working

as the Special Areas in 1934 (Central Scotland, North-east England, South Wales, and West Cumberland). Although this was not the first time that collieries and ironworks had closed in such localities, the scale and extent of dereliction was now far greater.

Serious though the rapid creation of so much derelict land was, particularly as it was coupled with severe social problems in the afflicted areas, the inter-war period was also marked by the more rapid increment of potential dereliction. For example, there were still 1,017 active collieries in 1947, many of them large undertakings capable not only of high levels of coal production but also of causing extensive damage through subsidence and the deposition of waste. Quarrying also presented greater potential difficulties after 1920 with the introduction of more widespread mechanisation, which, among other things, permitted working at greater depth than hitherto. The china clay industry of Cornwall produced particularly intractable land-use problems by deepening its pits and increasing the height of its waste heaps. Some forms of mechanised shallow quarrying were, however, equally disruptive, particularly in their after-effects. The 'hill and dale' left by the excavation of ironstone, and the flooded workings on the sites of abandoned sand and gravel pits were important new forms of derelict land created after 1920.

The active plants of the mineral-based industries gave rise to less potential dereliction after 1920. Iron and steel works increasingly converted slag into tarred road metal instead of tipping it; the slag produced by basic steel furnaces was, as noted above, a valuable by-product commonly used as fertiliser. Non-ferrous metal smelting was virtually dependent on imported concentrates, the smelting of which produced little or no worthless residue. The heavy chemical industry, although partly based on a waste-producing process, was responsible for a major technical innovation whereby one of its principal raw materials, brine, could be abstracted without causing damage from subsidence. The inter-war period was not, therefore, one of unremitting growth of dereliction in its various forms. There were also other signs of improvement, notably in the first steps towards framing legislation designed to minimise the disruptive after-effects of opencast mineral working. Significantly the introduction of opencast coal mining in 1942 was accompanied by an obligation to restore land after working had ceased, in itself an indication of the way in which official policy towards the problems of dereliction was moving.

3(ii) 1948 onwards

Fig 12 depicts the general pattern of potentially derelict land of various types in 1948–54 based on an analysis of contemporary topographical maps. The map summarises the major elements of dereliction (extant, potential and partial) which had developed during the past 200 years or so, and illustrates the position before most of the major schemes of reclamation were undertaken (except those at opencast coal mines). The similarities with the map for 1890–1900 are obvious. In the coalfields the areas of potential dereliction had spread further across the concealed measures, but in many localities past dereliction on the most accessible outcrops had been wiped out by the advent of opencast working. In the haematite mining districts of West Cumberland and Furness there was extensive dereliction (the last-named area ceased production in 1944), and these localities joined the non-ferrous metalliferous orefields as major tracts of extant derelict land. Slate quarrying had also contracted sharply (Fig 6) but several of the larger workings survived, producing waste which in extreme cases accounted for over 90 per cent of the mineral extracted.

There were, however, less predictable features. In 1948 the mineral-based industries had become insignificant sources of potential dereliction. On the other hand the mechanised quarrying of ironstone, sand, gravel, china clay, chalk and several other non-metalliferous minerals had given rise to much disruption, frequently in localities barely touched by dereliction at the turn of the century.

OTHER FORMS OF DERELICT LAND

Although much derelict land is created by extractive and mineral-based industries, there are several other major sources of dereliction. Many manufacturing industries produce solid waste which cannot be recycled or put to profitable use, but this can commonly be handled by the forms of controlled refuse disposal normally applied to towns' waste. However, in some instances the tradition of dumping waste haphazardly in the vicinity of the factory has persisted, as, for example, in the tipping of shraff (earthenware waste) at some potteries. More serious is the fact that certain processes in the engineering and metal-processing industries yield toxic wastes, and these, together with many of the residues from the plastics industry, present severe problems of disposal. It is possible that difficulties of this kind will increase if factories are obliged to purify liquid effluent before discharge, for the solids removed in the cleansing process will have to be disposed of elsewhere. A further problem stems from the increasing use of plastic components in consumer goods, for unlike metals and natural fibres these have little value as scrap and must be disposed of as solid waste. Unfortunately the existence of derelict land frequently attracts illicit dumping of solid waste as a cheap alternative to properly controlled disposal. Not only does this add to the unsightliness of derelict land, it also increases the cost of reclamation, which must ultimately be borne by the community at large. Even where derelict land has been reclaimed for use as public open spaces, local authorities commonly find that illicit waste tipping continues.

Disposal of solid waste from boiler furnaces has always been associated with steam-powered industrial plants, although it has rarely produced large-scale difficulties. Paradoxically the widespread use of electricity from the grid, often to replace individual steam engines at factories, gave rise to an exaggerated form of this problem in the 1950s, when it became necessary to

'age 67
ibove) Callow waste left by
iechanical excavation of Oxford
'lay for brick manufacture at
‹empston Hardwick,
ßedfordshire
below) Flooded workings in the
)xford Clay at Stewartby,
ßedfordshire. 'Stewartby Lake'
s to form the centrepiece of a
ountry park

Page 68 (*above*) The derelict site of Cyfarthfa ironworks, Merthyr Tydfil, 1928. The site was reclaimed for factory construction after 1937; (*below*) Dowlais, Merthyr Tydfil, 1960. The area at bottom right forms part of the cleared site of the Dowlais ironworks and its slag banks. Beyond the arterial road (*top*) the hillsides are broken by abandoned nineteenth-century coal and ironstone workings

dispose of increasingly large quantities of pulverised fuel ash produced at thermal generating stations. Fortunately in this, as in other instances, solutions were found. Pulverised fuel ash has been used as a raw material by the building and road construction industries, notably as aggregate, thereby reducing pressure on other mineral resources. In addition it has proved possible to reclaim the dumps of ash by direct treatment without the need to import top soil or to use other costly intermediates (see pp 247–8).

Disused industrial buildings and plant constitute another form of dereliction. The worst problems are associated with heavy industry employing specialised plant, for abandoned sites of this kind are unlikely to attract other users unless extensive remedial work has been carried out. An ironworks with its massive blast furnace foundations obviously presents a greater challenge to a would-be occupier than a textile mill or an engineering shed. However, increasing numbers of empty buildings, notably textile mills, have become derelict in localities which have little other than these obsolete premises to offer new enterprises. Abandoned military installations create a similar problem of dereliction unrelated to the pattern of mineral working. Disused military airfields formerly constituted a substantial proportion of this land, but many have been restored to agriculture or converted to other uses with little difficulty. Hutted encampments present few difficulties if they can readily be converted into sites for new housing and industry, as has frequently happened near towns. The remoter rural camps have been more difficult to clear, but increased demand for recreational facilities has led to their reclamation as caravan sites and other amenities. Even so, abandoned military installations may still form a high proportion of the derelict land in a locality. For example, a survey made in Cheshire during 1971 revealed that 29 per cent of the derelict land in the county comprised former army camps.

Abandoned transport installations form another major category of derelict land, although, with the exception of the

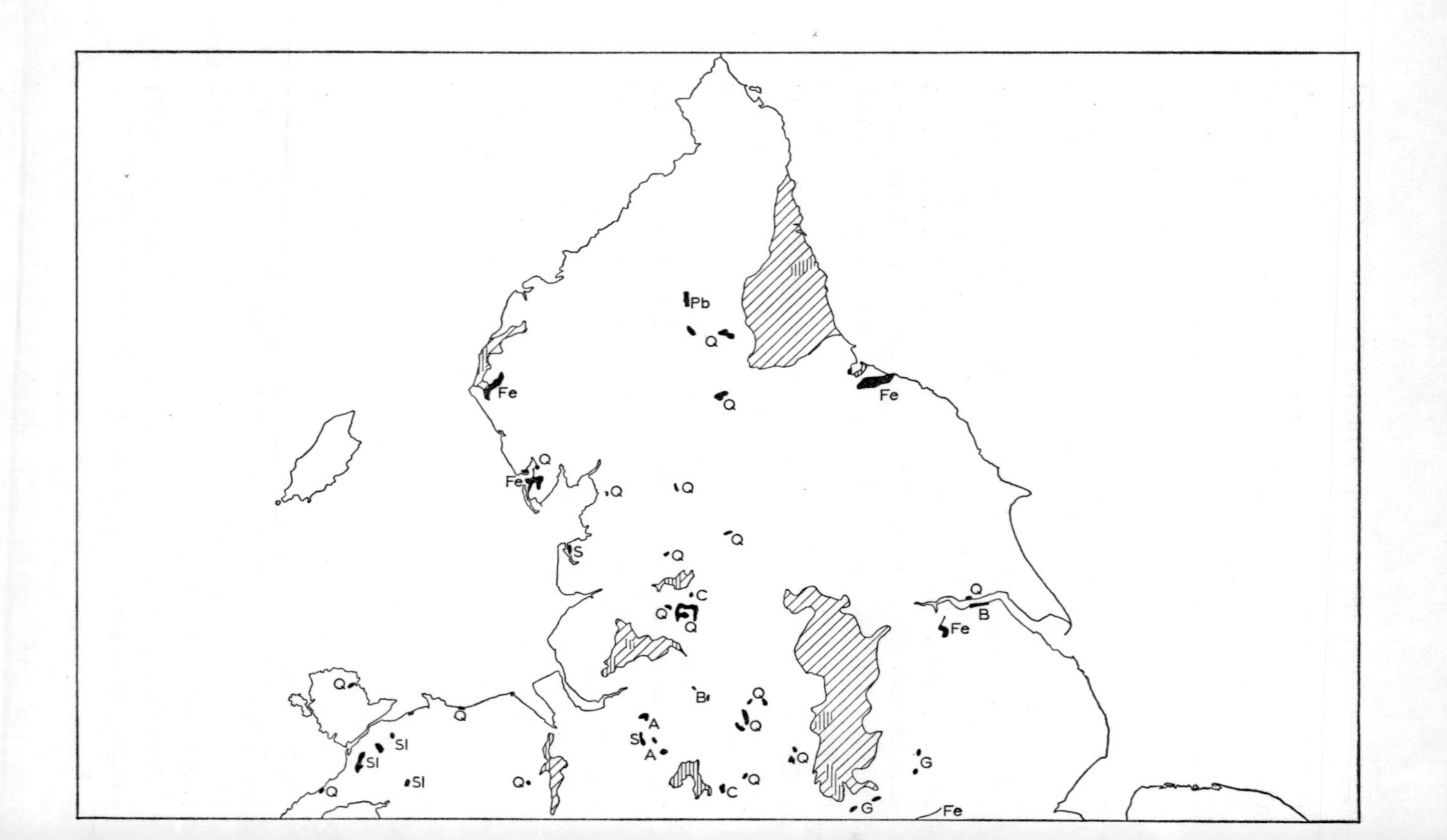
Pb
Q
Fe
Q
Fe
Q
Fe
Q
Q
Q
S
Q
C
Q
Q
Q
B
Fe
Q
B
Q
Q
A
Sl
A
Q
Q
Sl
Sl
Sl
Q
Q
Q
C
Q
G
G
Fe

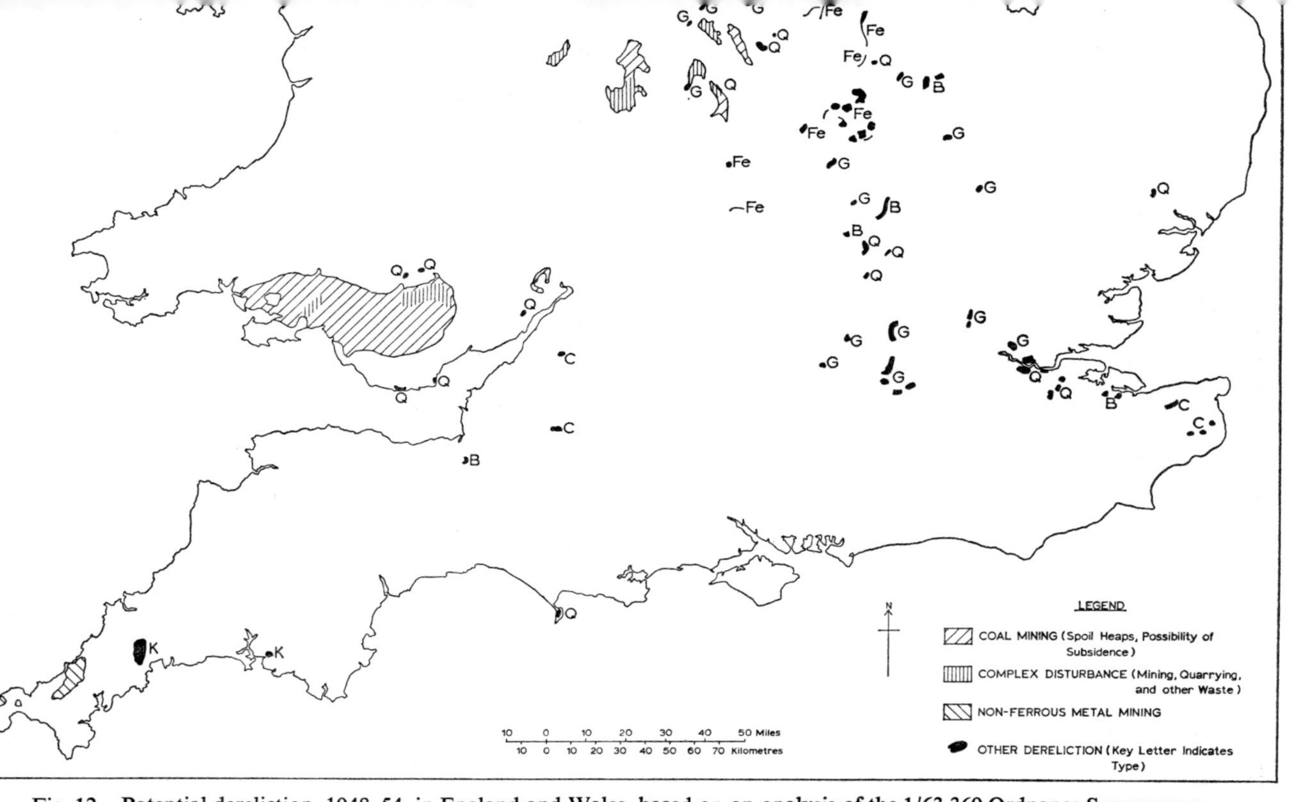

Fig 12 Potential dereliction, 1948–54, in England and Wales, based on an analysis of the 1/63,360 Ordnance Survey maps. Key as for Fig 11. G, gravel and sand working

railways, this aspect of dereliction has not been investigated on a national scale. Much of the dereliction stems from recent developments in transport technology. For example, changes in shipping practice, illustrated by the growth of container traffic, have caused extensive abandonment of obsolete dock installations in general cargo ports, notably London, whereas before the 1960s closures were largely confined to specialised ports, such as those handling exports of coal. Canal closures have gained momentum in the period since 1950, although in many instances the formal abandonment had been preceded by many years of neglect. Railway closures sincer 1948 have produced the most widespread dereliction in this category. In the years 1948–70, 35,000km of railway (excluding privately owned mineral lines) were closed, creating at least 87,500ha of derelict land, not all of which remained derelict in 1971. Derelict transport installations present a variety of challenging land-use problems, particularly in rural areas, but increased demand for recreational facilities has revealed the possibility of reclaiming many sections of abandoned canals and railways (see pp 248–51).

CONCLUSION

In 1948 the first estimate of dereliction in England and Wales suggested that there were 48,500ha of derelict land; in 1969 the official record indicated a total of 46,411ha. The detailed distribution of current dereliction is discussed in Chapter 3, but three general observations may be made at this point. Firstly, the continued growth of derelict land in the post-war period has been partly offset by effective reclamation schemes. Secondly, controls and legislation designed to minimise the creation of further derelict land have undoubtedly reduced the pace of dereliction, but they have only been a qualified success, particularly in established mineral working districts. Thirdly, the problems of dereliction have been greatly exacerbated by the abandonment of military installations, the closure of railways and canals and the accelerated rate of contraction in several mineral-working and mineral-using industries. However, it should not be thought that the absence of this factor would have made dereliction an insignificant feature of the changing pattern of land use. Estimates suggest that between 1,200ha and 1,500ha of land per year are lost through mineral working, in addition to which a further 4,600ha may be temporarily affected by the disruption which accompanies mining and quarrying.

The creation of derelict land is, therefore, a continuing problem which, on the best evidence available, still exceeds the annual amount of reclamation by about 300ha. Although much of the hard core of derelict land can clearly be attributed to the recent past, as an examination of Figs 11 and 12 suggests, it is unrealistic to suppose that earlier forms of land misuse are entirely to blame. Fortunately there is a growing awareness of the possibilities offered by schemes of land reclamation in derelict areas, not only from the standpoint of immediate financial returns but increasingly in relation to the wider socio-economic benefits which they may produce. It is equally significant that much of the machinery which is employed to cause the turmoil associated with mineral working can be as effectively employed in schemes of reclamation or, if proper planning is carried out, in the immediate restoration of land as quarrying progresses. Similarly methods of reducing mining subsidence and of minimising the ill-effects of surface waste tipping are technically feasible. These points are further examined in the concluding section of this book, but it is fitting to record in this history of the development of derelict land that major changes in attitudes to land use seem likely to reverse the trend of the past 200 years or so during the present decade.

References to this chapter begin on p. 309

CHAPTER THREE

The Distribution of Derelict Land, 1964–71

THE first official survey of derelict land in England and Wales took place in 1954 and recorded 51,274ha of dereliction. The data were collected by local authorities for the Ministry of Housing and Local Government, which later observed that they provided 'a rough and provisional estimate' of the problem, there being 'no comprehensive and uniform statistics of the extent and geographical distribution of derelict land in England and Wales' (MHLG 1956). The Ministry intended to improve its range of information by having data on derelict land included in the revisions of local authority Development Plans, but this did not secure an adequate flow of information. Consequently a separate annual survey of derelict land was initiated in 1964 and continued until 1971, when it was replaced by a more elaborate survey. The nature of the official returns has been discussed above (pp 34–6). In spite of their limitations they provide a consistently valuable source of information, and the basis of the analysis which follows.

The results of the surveys carried out between 1964 and 1971 are summarised in Table 3. During this period the amount of dereliction in England rose from 34,358ha to 39,292ha, an increase of 14 per cent, in spite of the fact that 6,216ha had been reclaimed during the same period. Although part of the increase is attributable to greater accuracy of survey, the successive annual returns clearly reveal that derelict land continues to grow more rapidly than reclamation can restore earlier dereliction to beneficial use.

TABLE 3 Summary of derelict land in England and Wales, 1964–71

Year	*Extent of derelict land in hectares*	*Year*	*Extent of derelict land in hectares*
1964	40,101 (34,358*)	1968	45,715
1965	43,511	1969	46,411 (38,738*)
1966	44,957	1970	39,133*
1967	45,499	1971	39,292*

* England only: the totals for England in 1964 and 1969 are shown in brackets for purposes of comparison

REGIONAL DISTRIBUTION

Viewed on a national scale dereliction does not seem to be an exceptional land-use problem, for in the 1960s dereliction rarely accounted for more than 3ha per 1,000 of the total area of England and Wales. This compares with an estimated 107ha per 1,000 under all forms of urban-industrial land use in 1960–61 (Best 1968). Similarly the gross rate of increment of derelict land, about 2,000ha per annum in the 1960s, compares with transfers from agricultural land to urban-industrial uses of about 17,000ha per annum. However, a more detailed examination of the distribution of derelict land reveals important areal variations in the intensity of the problem. At the level of the Economic Planning Regions derelict land is very unevenly distributed, particularly in its density (Table 4).

TABLE 4a Derelict land in England and Wales, 1969, by Economic Planning Regions

Economic Planning Region	Derelict land (hectares)	Per cent of national total	ha per 000 of total area	ha per 000 of urbanised area*	ha per 000 of undeveloped area*†
North	8,344	18·0	4·3	59·8	18·8
Yorkshire/Humberside	4,728	10·2	3·0	29·1	6·5
East Midlands	3,522	7·6	3·4	25·1	4·4
East Anglia	1,341	2·9	0·9	15·1	1·4
South East	2,132	4·6	0·8	4·3	1·9
South West	7,652	16·5	3·3	39·8	8·3
West Midlands	5,194	11·2	4·0	29·6	8·7
North West	5,825	12·5	7·3	28·0	27·2
Wales	7,673	16·5	3·7	55·3	8·6
England and Wales	46,411	100·0	3·1	26·8	7·1

Based on data from *Derelict Land Survey* and *Abstract of Regional Statistics* (1971)

* Based on estimates by R. H. Best in *Long Term Population Distribution in Great Britain*, Department of the Environment, 1971

† Total area minus urbanised area and that to which various planning constraints apply. All of this land is not necessarily capable of development

In the South East and East Anglia less than 1ha per 1,000 of the total area is derelict, compared with 4·3ha per 1,000 in the North and 7·3ha per 1,000 in the North West. Similarly large variations occur when dereliction is measured against the total area of urban land: the greatest difference is between the South East and East Anglia, with values of 4·3ha and 15·1ha per 1,000 respectively, and the North and Wales with values of 59·8ha and 55·3ha per 1,000 respectively. Finally, the relation between the recorded amount of dereliction and the balance of land remaining in each region, once the developed area and those localities subject to major planning constraints have been subtracted from the regional totals, shows that again the South East (1·9ha per 1,000) and East Anglia (1·4ha per 1,000) have the lowest ratios, while the North West (27·2ha per 1,000) and the North (18·8ha per 1,000) have the highest. Table 4a emphasises the fact that in southern England the reclamation of derelict land would provide very little relief in an area which has excessively heavy pressure on available living space. Table 4b outlines the regional composition of derelict land, using the threefold classification of the official statistics. In England and Wales the largest category is spoil heaps (42·3 per cent of all derelict land), followed by the heterogeneous category 'other forms of dereliction' (34·4 per cent) and excavations and pits (23·3 per cent).

TABLE 4b Composition of derelict land in England and Wales, 1969 by Economic Planning Regions

Economic Planning Region	Percentage of regional total			Percentage of national total		
	Spoil heaps	*Excavations and pits*	*Other forms of dereliction*	*Spoil heaps*	*Excavations and pits*	*Other forms of dereliction*
North	30·0	18·0	52·0	13·0	14·0	27·0
Yorkshire/ Humberside	24·0	42·0	34·0	6·0	18·5	10·0
East Midlands	27·0	31·5	41·5	5·0	10·0	9·0
East Anglia	—	36·0	64·0	—	4·5	5·0
South East	3·0	69·0	28·0	—	13·5	4·0
South West	75·0	10·0	15·0	29·0	7·0	7·0
West Midlands	38·0	16·0	46·0	10·0	8·0	15·0
North West	34·5	25·0	40·5	10·0	13·5	15·0
Wales	68·5	15·5	16·0	27·0	11·0	8·0
England and Wales	42·0	23·0	35·0	100·0	100·0	100·0

Based on data from *Derelict Land Survey;* values have been rounded off to the nearest 0·5 per cent

Exceptionally high proportions in a single category were recorded in the South West and Wales, where spoil heaps formed 75 per cent and 68·5 per cent respectively of the total dereliction; the South East, where pits and excavations formed 69 per cent of the total; and East Anglia, where 64 per cent of the total fell into the 'others' category. The last-named category was also the largest in four other regions (North, North West, East Midlands and West Midlands), surprisingly so as each contains a major coalfield in which spoil heaps might have been expected to form the bulk of the dereliction. It is a weakness of the statistics that they do not more adequately distinguish the character of 'other forms of dereliction', although it is possible to identify some of the principal elements by relating the figures for local authority areas to field evidence.

The composition of derelict land can also be examined in terms of the share of each category located within a given region as a proportion of the national total in that category. In the South West and Wales spoil heaps formed 29 per cent and 27 per cent respectively of the national total in this category. But pits and excavations in the South East and 'other forms of dereliction' in East Anglia were low as a percentage of the national total, since in both instances the high regional concentrations formed part of a relatively low total area of derelict land. Consequently it was the North, North West and West Midlands regions which had the largest share of land in the 'others' category, and Yorkshire-Humberside which contained the greatest share of the derelict pits and excavations.

COUNTY DISTRIBUTION

The data for individual counties provide the next level of areal refinement. For most purposes of analysis the returns for administrative counties can conveniently be aggregated with those for their associated county boroughs, but there is an important exception. In compiling Figs 13 and 14 data for administrative counties alone have been employed, as the subsequent analysis of notional costs of dereliction would be seriously distorted by the inclusion of rateable values for county boroughs. In fact the pattern of derelict land shown on Fig 13b is the same whether the statistics for county boroughs are used or not, for, although their inclusion produces a slight re-ordering of counties within the four categories, no county falls into a different category on this account.

In 1969–70 fourteen of the fifty-nine administrative counties in England and Wales had densities of dereliction higher than the national average of 3ha per 1,000: together they contained

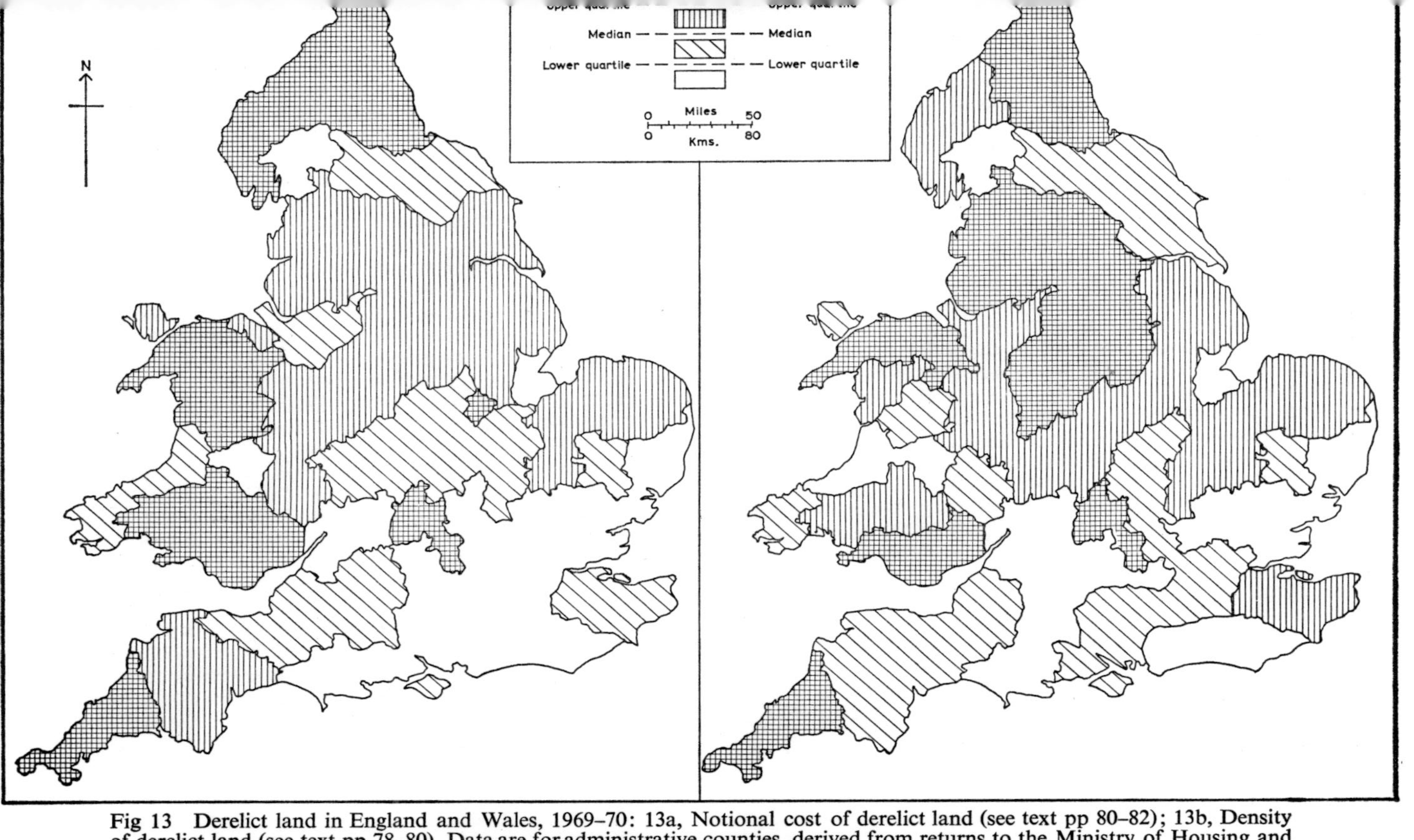

Fig 13 Derelict land in England and Wales, 1969–70: 13a, Notional cost of derelict land (see text pp 80–82); 13b, Density of derelict land (see text pp 78–80). Data are for administrative counties, derived from returns to the Ministry of Housing and Local Government and the Welsh Office

75 per cent of the derelict land but only 25 per cent of the total area of the administrative counties. The localities are identified in Fig 13b by the highest category (above the upper quartile, which is roughly coincident with the national average), and they range from Denbighshire (3·02ha per 1,000) to Cornwall (18·48ha per 1,000). The areas of highest density of dereliction correspond largely to the counties containing the major coalfields of Wales and of northern and midland England, together with North Wales, Cornwall and Oxfordshire. Cornwall and North Wales qualify for inclusion mainly through the presence of various types of mineral waste heap and abandoned quarries. Oxfordshire (3·70ha per 1,000), a less expected member of this group, includes in its 718ha of derelict land many abandoned gravel workings.

The counties with moderate densities of dereliction (between the median and the upper quartile) fall into two groups. The largest comprises localities peripheral to and sharing many characteristics of the high density areas, notably in midland England and south-central Wales. The second group lies mainly in eastern England, where much of the dereliction comprises disused transport installations and abandoned surface mineral workings. In Kent, which forms part of this group, coalfield dereliction is added to these two forms of derelict land. In the remaining counties, below the median value of 1ha per 1,000, almost the whole of southern England displays low densities of derelict land, together with much of mid-Wales and a belt of country running from the Lake District to the Yorkshire coast.

Although the areal density of dereliction provides one measure of spatial variation, it would be useful to measure these differences on a cost basis. There are, however, few sources of data on the costs of dereliction to the community at large, for even though many counties are currently preparing estimates of reclamation costs in order to mount a major attack on derelict land during the 1970s, this information is not universally available. It is possible to calculate the notional cost of dereliction in each county by using its rateable value and the official statistics of dereliction. The calculations are based on two assumptions: firstly, that all recorded derelict land merits reclamation at a cost equal to the national average of approximately £3,200 per hectare and, secondly, that the expense has to be borne by local authorities through a levy on the rates (local taxes). Although this last point does not accord with actual practice (see pp 215–17), there is evidence to suggest that the poorer local authorities find difficulty in bridging the financial gap between state grants in aid and the total cost of reclamation, so that the exercise is not entirely divorced from reality.

The notional cost of reclamation in a local authority area is calculated by expressing the cost of reclamation (amount of recorded derelict land multiplied by the national average gross cost of reclamation) as a percentage of its total rateable value. In effect this gives the levy in pence per pound of rateable value were a local authority to fund the cost of reclamation from its own finances in a single year. Although this is again far from being current practice, each local authority has to bear part of the expense, so that the actual cost entailed will be a reduced proportion of the notional cost. Evidence taken from estimates produced by several counties suggests that the actual cost to local funds is from 10 to 25 per cent of the notional cost cited below. This range is largely explained by local variations in the level of state grants in aid and differences in the net costs of actual reclamation schemes.

The notional cost of dereliction in each administrative county is shown in Fig 13a. The highest category, with notional costs above 21 per cent of the total rateable value, comprises a variety of counties. Cornwall again leads (160 per cent), and much of mid-Wales ranks high in this category. The coalfield counties of South Wales and the North region, together with Rutland and Oxfordshire, form the remaining high-cost areas. The counties with moderate notional costs (between 8 and 21 per cent of rateable valuation) form a broad swathe covering much of eastern England, the northern Midlands, Lancashire and Yorkshire, together with the periphery of high-cost areas in North Wales and South West England. The principal low-cost tract is in southern England where, in the South East region particularly, notional costs of dereliction commonly fall to less than 0·5 per cent of rateable valuation.

In Fig 14 the data relating to density of dereliction and notional costs expressed as rateable burdens have been amalgamated to produce five categories of combined density and cost of derelict land. The highest category, with both sets of values above the upper quartile, depicts the principal problem areas, with intense dereliction and high rate burdens. The coalfield counties of North East England and South Wales, and the mining and quarrying districts of North Wales and Cornwall, form the major components, together with Oxfordshire, which marginally qualifies for inclusion on both scores. The second category, with both values above the median (one of which may also be above the upper quartile) includes much of eastern England and counties peripheral to the two major Welsh concentrations of dereliction, all localities with relatively low rural rateable values. In addition the remaining major coalfield counties fall into this category, where they form two distinctive subdivisions: the first comprises the counties which combine a moderate density of dereliction with high rate burdens that are a reflection of poor rateable values, eg Cumberland and Carmarthenshire; and the second comprises the counties with high density of dereliction and low rate burdens that are based on high rateable values, eg Lancashire and Staffordshire. The same kind of distinction separates the next two categories. Those counties with derelict land densities below but rate burdens above the median are dominantly rural with low total rateable valuations. Those with derelict land densities above the median but rate burdens below are relatively wealthy counties. Worcestershire, Warwickshire and Leicestershire form a transitional zone of this type between the highly burdened north and the lowly burdened south east. More noticeable are the positions of

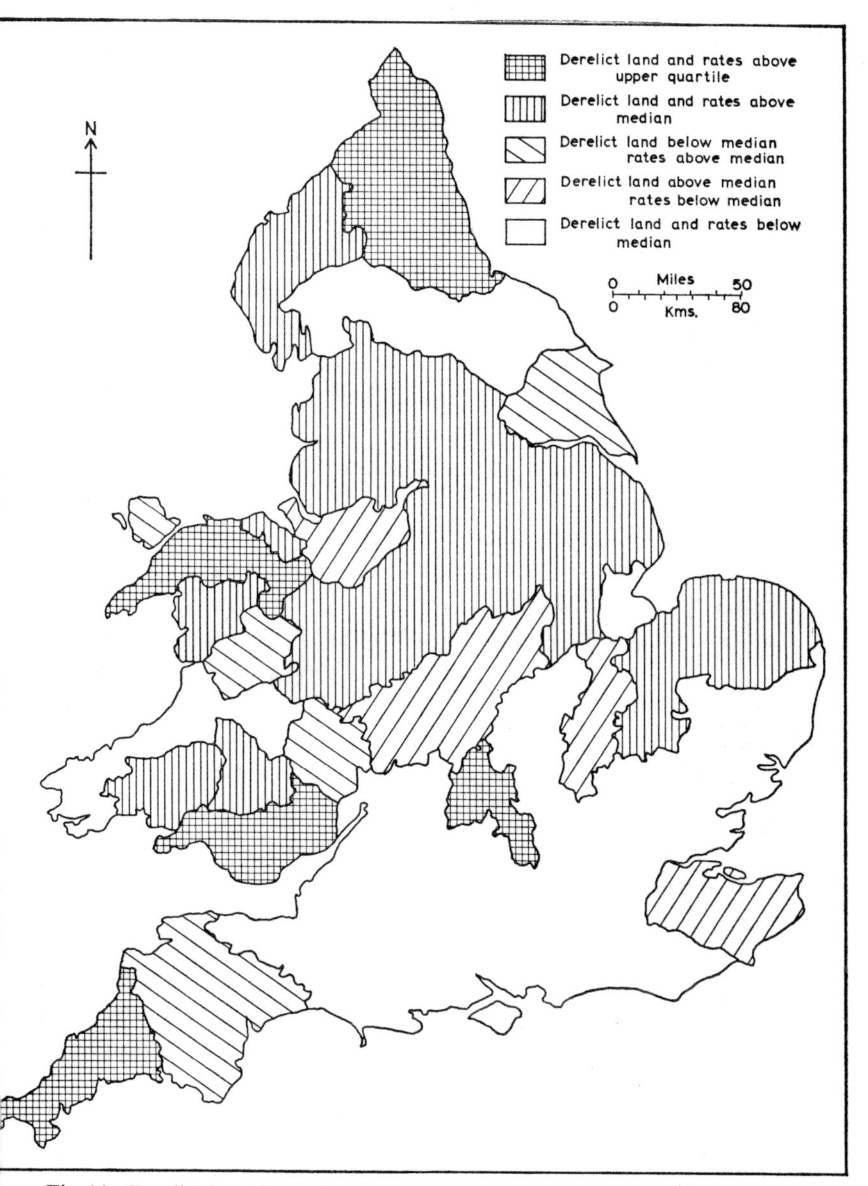

Fig 14 Derelict land in England and Wales, 1969–70. In this map the data employed in Figs 13a and 13b have been combined (see text pp 82–4). The notional cost of deduction is expressed in terms of rateable burden (abbreviated to 'rates' in the legend).

Cheshire in the north and Kent in the south-east as moderate-density low-burden enclaves. The fifth category comprises the counties that have the fewest problems of dereliction, with both values below the median. Almost the whole of southern England falls into this category, together with several dominantly rural counties in mid-Wales and northern and eastern England.

Fig 15 shows the amount and composition of derelict land in the counties and their associated county boroughs in 1969–70; in some instances the returns have been further aggregated to combine those for adjacent counties where the individual totals are small. The largest amount of dereliction is in Cornwall (6,489ha), and only three other counties have more than 4,000ha: Lancashire (5,151ha), Durham (4,367ha), and the West Riding of Yorkshire (4,209ha). The importance of the counties containing the major coalfields is clearly seen, for none recorded less than 500ha of derelict land. Conversely, only three non-coalfield counties recorded higher amounts: Caernarvonshire (839ha), Oxfordshire (718ha) and Norfolk (640ha). Cheshire (589ha) may also be considered to form part of this group, even though it contains the worked-out extension of the Lancashire coalfield. At the other extreme, several counties returned less than 100ha of derelict land, notably in southern England, and one, Hertfordshire, made a nil return.

The composition of derelict land in each county is mainly remarkable for the number of instances in which a single form of dereliction is dominant, with over 60 per cent of the recorded total attributed to it. Fifty-two counties made returns, excluding a further seven which made single-category returns of less than 20ha, and of these thirty-five contained one dominant category. Spoil heaps formed the dominant category in seven instances, commonly accounting for over 80 per cent of a county's derelict land, as in Shropshire (93 per cent) and Cornwall (85 per cent). In fifteen counties excavations and pits formed the dominant category, again with some very high values, as in Bedfordshire (96 per cent) and Essex (77 per cent). 'Other forms of dereliction' was the dominant category in thirteen counties, and yet again

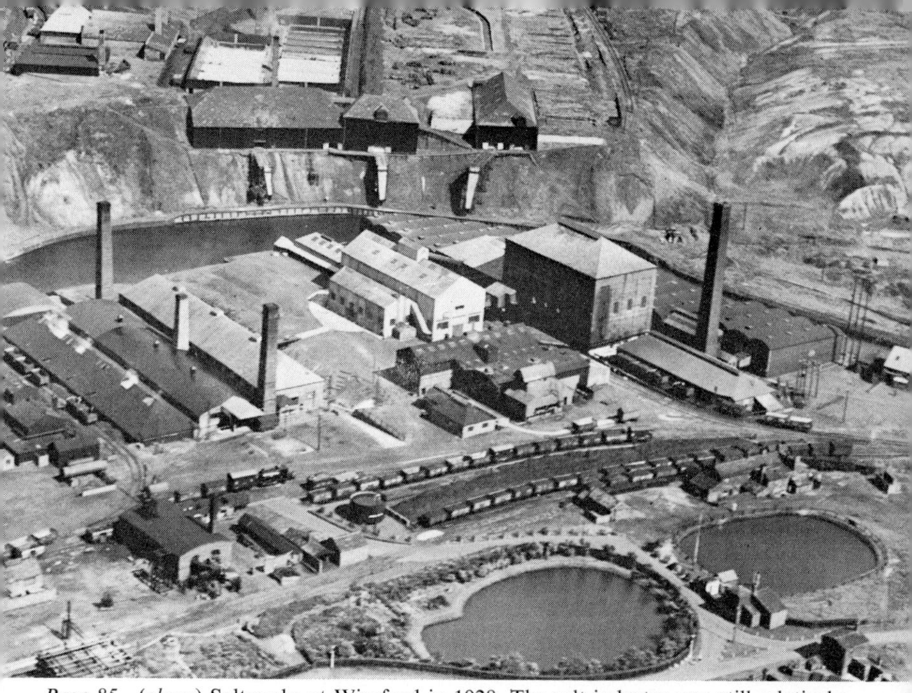

Page 85 (*above*) Saltworks at Winsford in 1928. The salt industry was still relatively prosperous at this time, but the sites of abandoned works could already be seen (*top centre*). The broken ground beyond and to the right of the ravine was partly the product of brine-pumping subsidence, partly used for dumping clinker from the open-pan furnaces. Part of this area appears in plate on p 175; (*below*) Derelict land near Northwich, Cheshire, 1971. The Winnington-Wallerscote ammonia-soda works complex occupies land on the left bank of the River Weaver. Beyond Wallerscote works (*top centre*) are the lime beds, on former agricultural land; disused lime beds in the floodplain above the works (*left centre*) occupy land damaged by subsidence. The strip of scrub-covered land on the right bank of the river comprises the sites of abandoned saltworks, some of which have also been used as waste dumps (see also plate, p 157)

Page 86 (*above*) Colliery spoil heaps at Pilsley, Derbyshire, together with an abandoned section of the Great Central Railway: photograph taken in 1969; (*below*) The same view as preceding plate, taken in 1972 after the completion of a major reclamation scheme

there were examples of nearly absolute dominance in Herefordshire (99 per cent), Hampshire (91 per cent) and Norfolk (89 per cent), in each case amounting to over 200ha of derelict land.

In the remaining seventeen counties no category of dereliction exceeded 60 per cent, although eight of these listed at least half their derelict land in a single category. Consequently only nine counties recorded a broad spectrum of dereliction (Table 5), and almost without exception these were localities with complexes of coal mining and other extractive industries, together with miscellaneous elements of economic decline reflected in the category 'other forms of dereliction'. The sole exception was Greater London, but with only 190ha of derelict land this cannot be considered a major deviation from the general point made above. An examination of the percentage values returned by all the counties for each category shows wide ranges of values which give rise to relatively high deviations from the mean; 20 per cent for the category spoil heaps, 26 per cent for excavations and pits, and 27 per cent for 'other forms of dereliction'. Thus it is not only the distribution of derelict land which displays important spatial variations but also the composition of dereliction.

TABLE 5 Derelict land, 1969–70: counties with no category of dereliction exceeding 50 per cent of the total

			Percentage in each category: deviation from national average in brackets		
County	*Derelict land (ha)*	*Derelict land per 000ha*	*Spoil heaps*	*Excavations and pits*	*Other forms of dereliction*
Flintshire	325	4·9	35 (—7)	38 (+15)	27 (—8)
Denbighshire	520	3·0	45 (+3)	24 (+1)	31 (—4)
Lancashire	5,151	10·7	36 (—6)	24 (+1)	40 (+5)
Cumberland	910	2·3	32 (—10)	36 (+13)	32 (—3)
Yorkshire WR	4,209	5·8	30 (—12)	35 (+12)	35 (=)
Derbyshire	1,696	6·6	43 (+1)	12 (—11)	45 (+10)
Nottinghamshire	1,073	4·9	28 (—14)	45 (+22)	27 (—8)
Staffordshire	3,073	10·4	36 (—6)	19 (—4)	45 (+10)
Greater London	190	1·2	22 (—20)	32 (+9)	46 (+11)

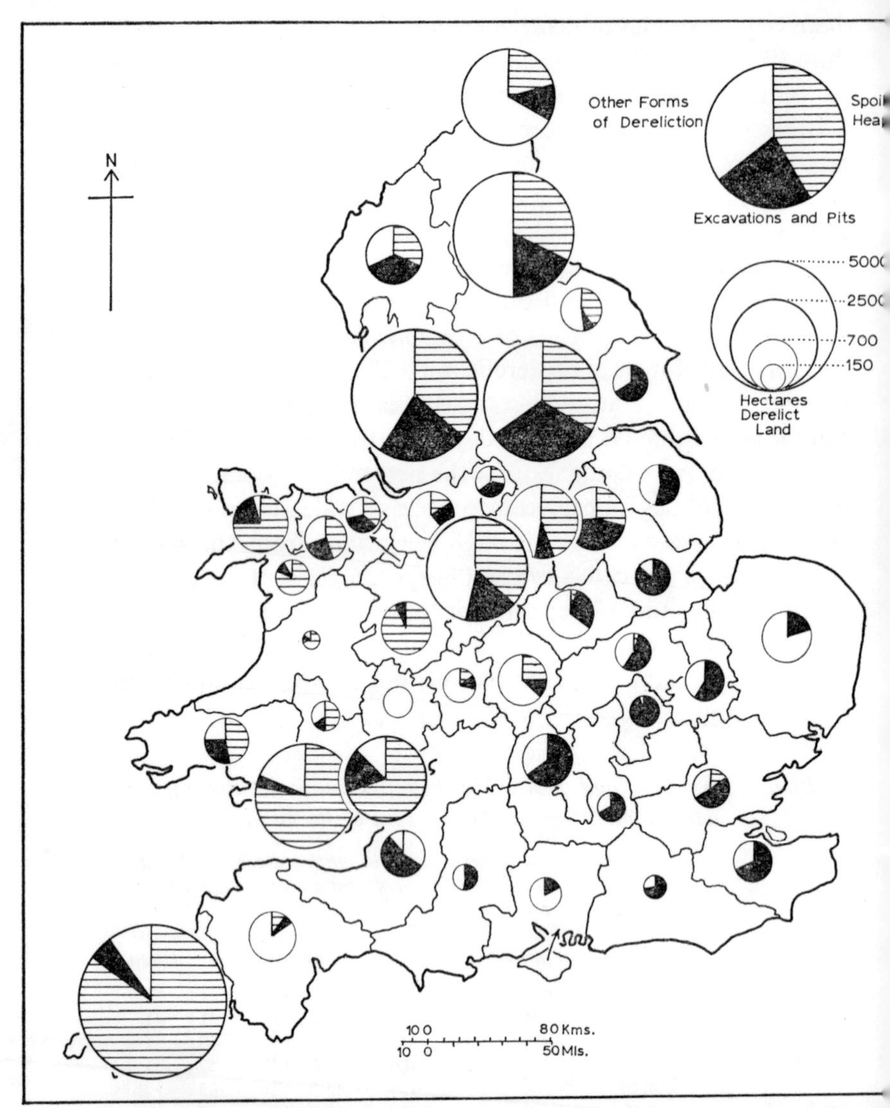

Fig 15 Composition of derelict land, 1969–70. Total amount of dereliction, distinguishing three categories of derelict land, by administrative counties and associated county boroughs (in those instances where counties recorded small amounts of dereliction the returns for adjacent areas have been amalgamated). Data are from returns to the Ministry of Housing and Local Government and the Welsh Office

LOCAL DISTRIBUTION

The smallest unit areas for which the official statistics are available are those administered by local authorities of different status. For purposes of physical planning, including the reclamation of derelict land, most local authorities are subservient to the administrative counties. The major exceptions are the county boroughs, which have parity of status with the administrative counties and, therefore, formulate their own planning policies. They merit separate attention for this reason, and also because they were deliberately omitted from the section dealing with density and notional costs of dereliction on pp 78–84. (All the references to administrative areas in this chapter relate to the local government structure before reorganisation on 1 April 1974.) In 1969 the English county boroughs contained 4,543ha of derelict land, roughly 12 per cent of the national total. (Comparable figures are not available for Wales.) The median density was 5·42ha per 1,000, with a lower quartile composed of nil returns and an upper quartile value of 15·02ha per 1,000. All the county boroughs with values above the median were situated in northern and midland England, while those with values above the upper quartile were almost exclusively located in Lancashire, the West Riding of Yorkshire, and Staffordshire (Fig 16). Consequently the pattern of density in the county boroughs was very largely complementary to that in the administrative counties, with the exception of North East England and Cornwall, the last-named having no towns of county borough status.

The notional costs of dereliction in the county boroughs were exceedingly low when compared with those for other local authority areas in the same district, even in the localities recording high densities of dereliction. The median value was 0·8 per cent of rateable value, and the upper-quartile value was 3·0 per cent, providing evidence of the great concentrations of rateable wealth in most county boroughs. The county boroughs

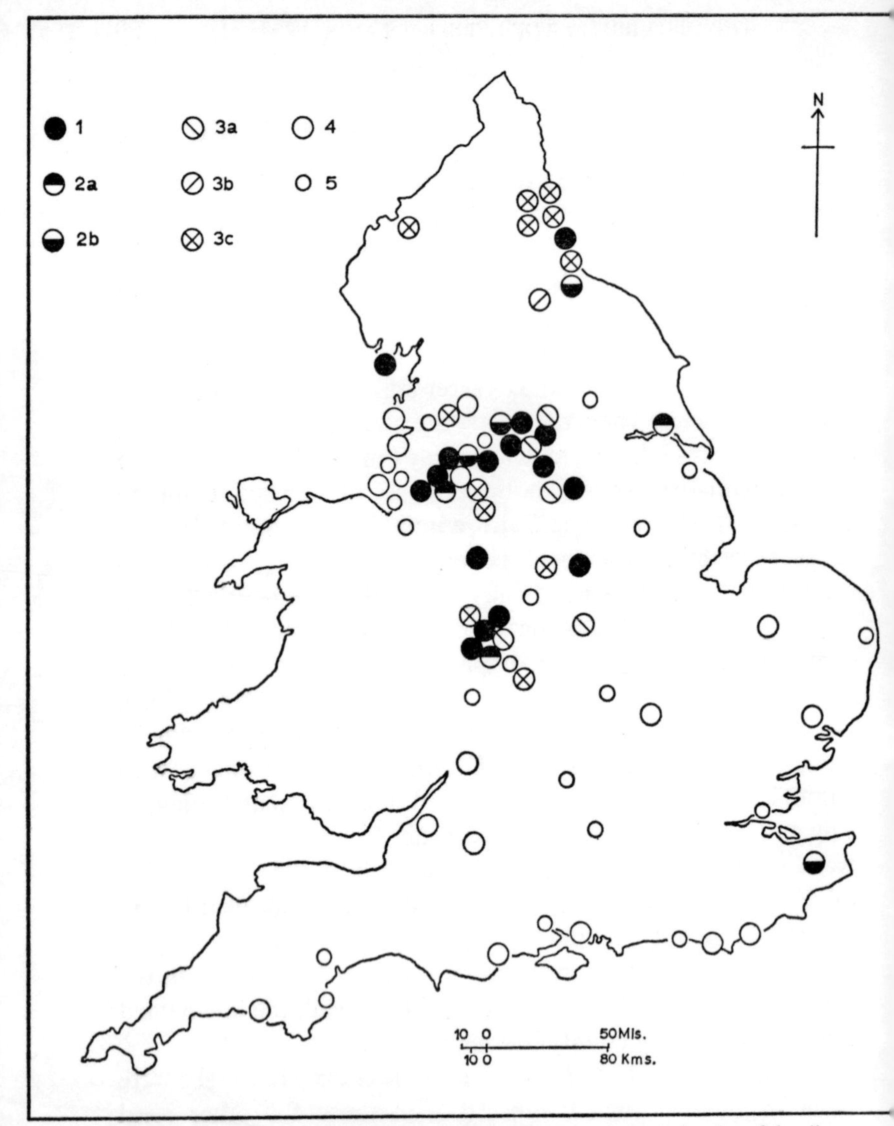

Fig 16 Derelict land, 1969, English county boroughs: 1, both density of dereliction and notional cost above upper quartile; 2a, density above upper quartile, notional cost above median; 2b, density above median, notional cost above upper quartile; 3a, density above median notional cost below; 3b, notional cost above median, density below; 3c, both density and notional cost above median; 4, both density and notional cost below median; 5, nil return. (For values in each instance see text pp 89–91.) Data are from returns to the Ministry of Housing and Local Government

recording values above the median were, with one exception, situated in northern and midland England, and similarly those recording values above the upper quartile were in Lancashire, the West Riding of Yorkshire and Staffordshire. The exception was Canterbury CB, with a notional burden of 4·0 per cent on the lowest total rateable value of any English county borough. The highest burden recorded was that for Barrow in Furness CB (24·3 per cent), followed by Stoke on Trent CB (20·1 per cent), Dudley CB (17·6 per cent) and Barnsley CB (16·9 per cent).

The areal extent of derelict land and its composition in the county boroughs have already been considered in general terms by aggregating their figures with those for their associated administrative counties (pp 84–7), and they are further discussed for selected areas in Chapter 9. The composition of derelict land in the county boroughs comprised 1,352ha of spoil heaps (30 per cent of the total), 1,057ha of excavations and pits (23 per cent), and 2,134ha of 'other forms of dereliction' (47 per cent). Almost half the spoil heaps were located in the Staffordshire county boroughs, with 227ha in Stoke on Trent alone. Excavations and pits were more widely distributed, with particularly large areas in Stoke on Trent CB (138ha), Barrow in Furness CB (121ha) and Walsall CB (100ha). Land in the category 'other forms of dereliction' was most widespread of all, but again the Staffordshire county boroughs contained a disproportionately large share (917ha, 43 per cent of the total). Not surprisingly, therefore, the Staffordshire county boroughs recorded the highest amounts and densities of derelict land: 667ha in Stoke on Trent CB (77·70ha per 1,000), 443ha in Dudley CB (73·48ha per 1,000), 301ha in Walsall CB (57·35ha per 1,000), and 250ha in West Bromwich CB (52·63ha per 1,000).

The remaining local authority areas, not of county-borough status, range in size from the largest rural district (116,000ha) to the smallest municipal borough or urban district (34ha). The detailed analysis of returns for the 1,312 local authority areas in England and Wales has not been undertaken, but data for several of the localities with severe dereliction have been examined, and those for England and Wales as a whole have been employed in Fig 17, which provides a generalised impression of the distribution of derelict land. This map reinforces the picture derived from an examination of the county returns. Not surprisingly the localities with heavy dereliction (over 10ha per 1,000) are mainly to be found in those counties already identified as the major problem areas. There are, however, pockets of heavy dereliction within counties which are not in this category, notably in the Cheshire saltfield, the Cleveland iron ore field, the gravel working districts of Greater London, and the Somerset coalfield. The areas with moderate dereliction (under 10ha per 1,000) are much more widespread and almost defy generalisation. Two points can, however, be made. Firstly, many rural districts fall into this class, and examination of several of these localities shows that they contain pockets of heavy dereliction which are masked in the returns by the great territorial extent of these local authority areas. Secondly, several counties falling into the two highest categories of density of dereliction (Fig 13b) contained no local authority areas with more than moderate dereliction: Oxfordshire, Bedfordshire and Norfolk are outstanding examples of this kind. Many local authorities returned no derelict land (Fig 17), notably in southern England, south-central Wales, the North Riding of Yorkshire and Lincolnshire.

Five examples have been chosen for a more detailed analysis at local authority level: Cornwall, Durham, Kent, Leicestershire and Northamptonshire. In addition similar data are also used in Chapter 9, which examines the development of derelict land over a period of time in selected areas.

Cornwall is an obvious choice for further study, for it heads the list in each of the measurements of dereliction employed above. Whether it would occupy the same unfortunate position in an objective assessment of the visual impact of derelict land is very doubtful, as the county enjoys a reputation for attractive scenery. In fact, its dereliction has often been described in a way normally reserved for more orthodox facets of scenic grandeur.

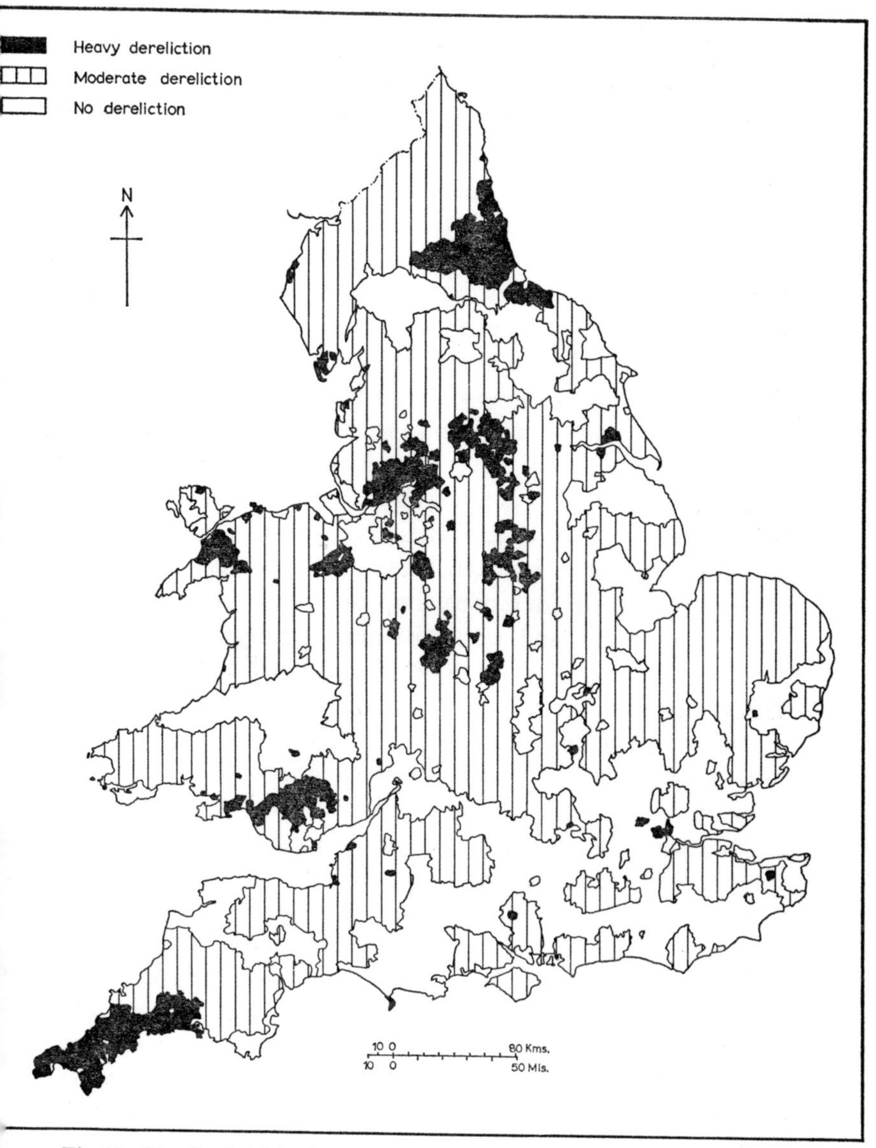

Fig 17 Derelict land in England and Wales, 1969, by local authority areas. Intense dereliction, over 10ha per 1000; moderate dereliction, under 10ha per 1000. Note that many local authorities returned no derelict land. Data are from returns to the Ministry of Housing and Local Government and the Welsh Office

Thus Daphne du Maurier (1967) writing of Wheal Cotes, St Agnes, observed that 'the headland . . . is wild and bare, save for the bleak walls of the engine-house, standing like a defaced cathedral against the sky, while beneath it a solitary chimney-stack formidable as a dark tower, frowns down upon the sea'. Nor is this romanticism confined to the long-abandoned tin and copper mines, for of the district around Hensbarrow Down, St Austell, it is noted that 'these clay heaps, with their attendant lakes and disused quarries, have the same grandeur as the mines in decay but in a wilder and more magical sense'. Few areas of derelict land have inspired such prose, and one is bound to agree that the uniqueness of the Cornish industrial scene diminishes the seriousness of the problem suggested by the crude statistics. The county planning officer has made the same point: 'In Cornwall derelict land is not seen as a blot on the landscape as it is in other parts of Great Britain because much of it has been colonised by gorse, bracken, heather and similar vegetation. Many feel that any attempt to "improve" such areas in the sense of reclaiming the land would be quite wrong even if it were feasible and reasonably economic' (personal communication, 1972). Nevertheless Cornwall's leading position commands attention, even if the area covered by derelict land is considered to be a false indicator of the severity of the problem.

In 1969 there were 6,489ha of derelict land in Cornwall. As Fig 18 shows, the majority of the county's local authority areas contained derelict land above the national upper-quartile value (3·00ha per 1,000), with particularly high levels in St Just UD (71·78ha per 1,000) and Camborne-Redruth UD (69·12ha per 1,000). Both localities contain active tin mines, but their derelict land is a product of the abandonment of earlier workings, mainly during the period 1870–1920. Much of it comprised low mounds of spoil, overgrown with gorse and heather, together with abandoned engine-houses and other pit-head buildings. In addition land close to the calciners, in which arsenical impurities were removed from the ore, was also made derelict by atmospheric pollution. Large amounts of derelict land were

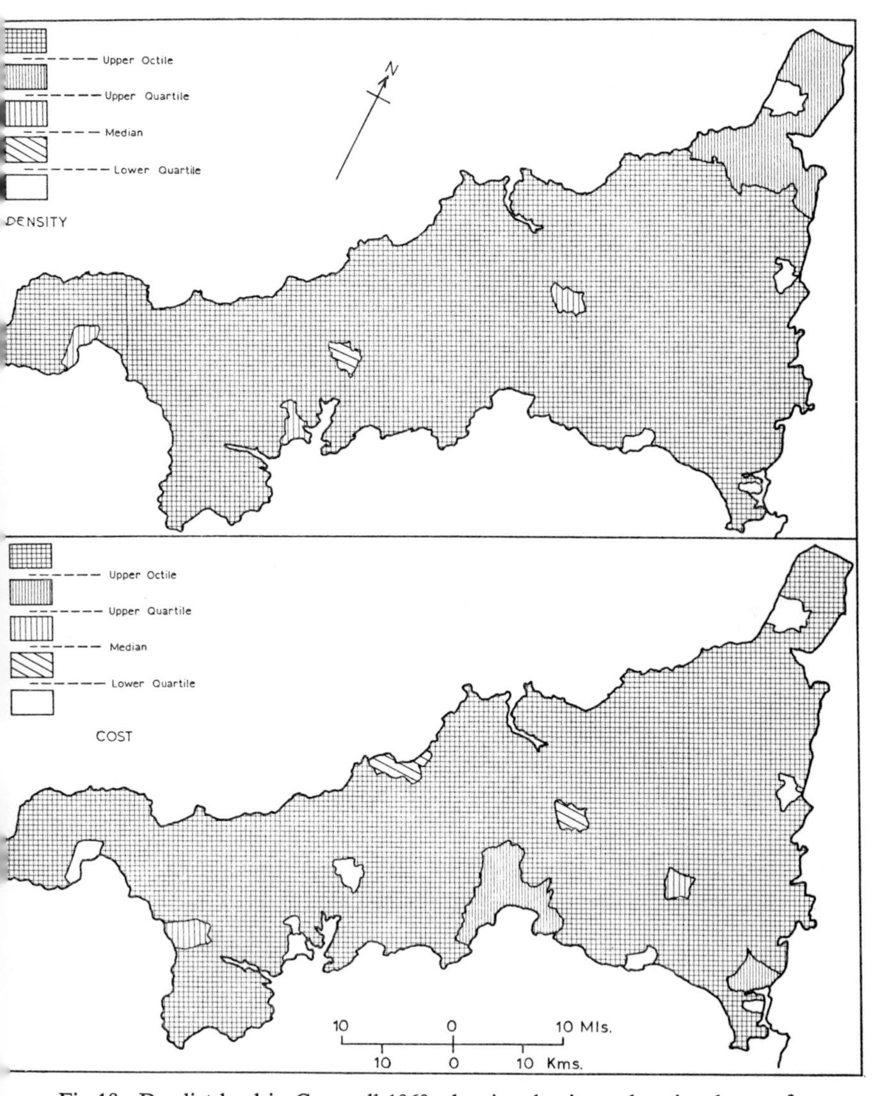

Fig 18 Derelict land in Cornwall 1969, showing density and notional cost of dereliction by local authority areas. NB: in this map and Figs 19–22 the five shading categories derive from national values used in Fig 13. Density: upper octile 5·1, upper quartile 3·0, median 1·0, lower quartile 0·4 (hectares per 1,000). Notional cost: upper octile 42, upper quartile 21, median 8, lower quartile 1·9 (per cent). Data are from returns to the Ministry of Housing and Local Government

also recorded by several rural districts, in which the densities appear to be relatively low owing to the great extent of these local authority areas. The largest amounts were in Truro RD (1,548ha, 35·31ha per 1,000), Kerrier RD (1,161ha, 31·58ha per 1,000), and West Penwith RD (741ha, 30·54ha per 1,000). Again most of this dereliction was related to mines and ancillary plant. The western part of Truro RD forms an extension of the major industrial area centred on Camborne-Redruth, including the derelict coastal mines around St Agnes, and the formerly important copper mining district of Gwennap-St Day, where output reached its peak in the mid-nineteenth century. The northern parishes of Kerrier RD form a similar extension of the Camborne-Redruth mining district, while West Penwith RD incorporates the coastal mining area between St Just and St Ives, together with the decayed port-cum-industrial centre of Hayle. The former non-ferrous mining area of West Cornwall, including the five leading localities named above, contained 4,122ha of derelict land in 1969 – two-thirds of the county's total dereliction situated in one-quarter of its total area.

The remaining derelict land in Cornwall lies in two major localities, although little of eastern Cornwall is free from dereliction. The adjacent rural districts of Launceston, Liskeard and St Germans contained 1,109ha of derelict land (12·10ha per 1,000), much of it again related to non-ferrous mineral extraction, and to a lesser extent china-clay working and granite quarrying. In the St Austell area there were 477ha of derelict land (11·37ha per 1,000) mainly related to china-clay working. In Cornwall as a whole spoil heaps were the dominant form of dereliction, and this is reflected in the fact that within the major concentrations identified above spoil heaps accounted for 87 per cent of the derelict land.

The notional costs of dereliction, expressed as a burden on the local rates, were very high in Cornwall generally, but they rose to exceptional heights in the poorer rural areas, notably Launceston RD (815 per cent), Kerrier RD (549 per cent), Truro RD (527 per cent) and Liskeard RD (519 per cent). St Just UD (in fact a largely rural locality) had a rateable burden of 715 per cent; other urban areas with high rateable burdens included Camborne-Redruth (159 per cent), St Ives (51·0 per cent), and St Austell-Fowey (34·0 per cent). Cornwall was, therefore, not only characterised by high densities of dereliction, it was also in an apparently weak position to remedy the problem from local funds. In spite of the fact that much of the dereliction may not warrant reclamation, certain mineral-working areas pose potential problems, and in consequence efforts are being directed towards mitigating the effects of current extraction rather than removing the residues left by earlier mining and quarrying.

County Durham ranks second to Cornwall among English counties in terms of both density and notional costs of dereliction. But no one has romanticised its coastal collieries, with their practice of tipping great tonnages of waste over the cliffs to darken the beaches below – although J. B. Priestley (1934) did liken the spoil heaps at Shotton Colliery to volcanoes! In Durham (Fig 19) the highest densities of dereliction were to be found along the south bank of the Tyne, in Jarrow MB (74·09ha per 1,000) and Felling UD (62·59ha per 1,000), and in much of West Durham, where the small Tow Law UD recorded the highest density of all (129·97ha per 1,000). The West Durham coalfield contained 51·5 per cent of the county's derelict land (Table 6), compared with 25 per cent in the East Durham coalfield and 20·5 per cent in the Pennine dales and uplands. In the last-named locality half the derelict land comprised abandoned quarries in Weardale, and a further third comprised spoil heaps, largely in the formerly important non-ferrous mining districts of the Pennine dales. The composition of derelict land in the two subdivisions of the coalfield was broadly similar. Somewhat surprisingly the category 'other forms of dereliction' was the most important, although in many instances it must have included abandoned colliery buildings and ancillaries, including mineral railways, together with land damaged by mining subsidence. The higher share of dereliction in West Durham is

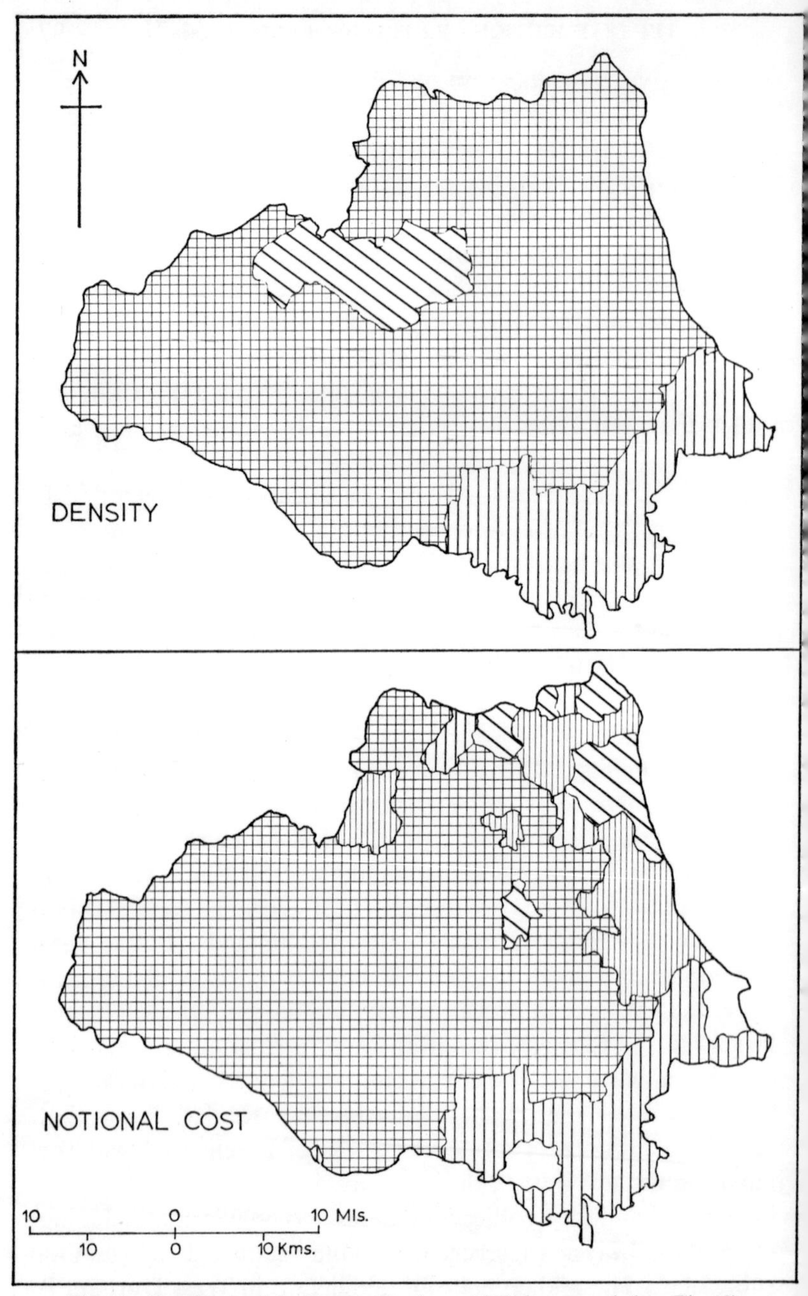

Fig 19 Derelict land in County Durham, 1969 (see note below Fig 18)

largely explicable in terms of its longer history of industrial decay. However, East Durham presents a greater potential problem, for here the pits are larger, and any acceleration in the NCB's programme of colliery closure would add greatly to the burden of dereliction. Along the coast, where most of the large surviving collieries are located, much spoil is tipped into the sea; although this reduces the amount of space consumed by waste heaps on land it does not provide a satisfactory means of reducing the volume of potential dereliction.

TABLE 6 Derelict land in County Durham, 1969, by major regions

*Region**	*Total derelict land (ha) and percentage share in county*		*ha derelict land per 000*	*Spoil heaps per cent*	*Excavations and pits per cent*	*Other per cent*
West Durham	2,177	(51·5)	28·21	33	10	57
East Durham	1,066	(25·0)	20·95	36	11	53
Pennine Uplands	867	(20·5)	10·22	33	50	17
Total†	4,110	(97·0)	19·31	34	19	47

* Based on entire local authority areas
† The remaining 3 per cent is located on Teesside

The notional costs of dereliction are shown in Fig 19. The heaviest rate burdens fall on the rural districts of the Pennines, with 471 per cent in Weardale RD, and 258 per cent in Barnard Castle RD. In West Durham the notional cost of dereliction averaged 45·8 per cent, but particularly high local burdens were recorded by Tow Law UD (158 per cent), Lanchester RD (146 per cent), and Crook and Willington UD (141 per cent). In East Durham the notional cost was substantially lower (15·8 per cent), and the highest recorded local value was 93·0 per cent in Sedgefield RD. This emphasises a fundamental difference between the two parts of County Durham which is evident in many ways, not least in the designation of the more prosperous eastern half as part of the North East's 'growth zone'.

Kent and Leicestershire (Figs 20–21) are taken as examples of counties which do not display the extremes exemplified by

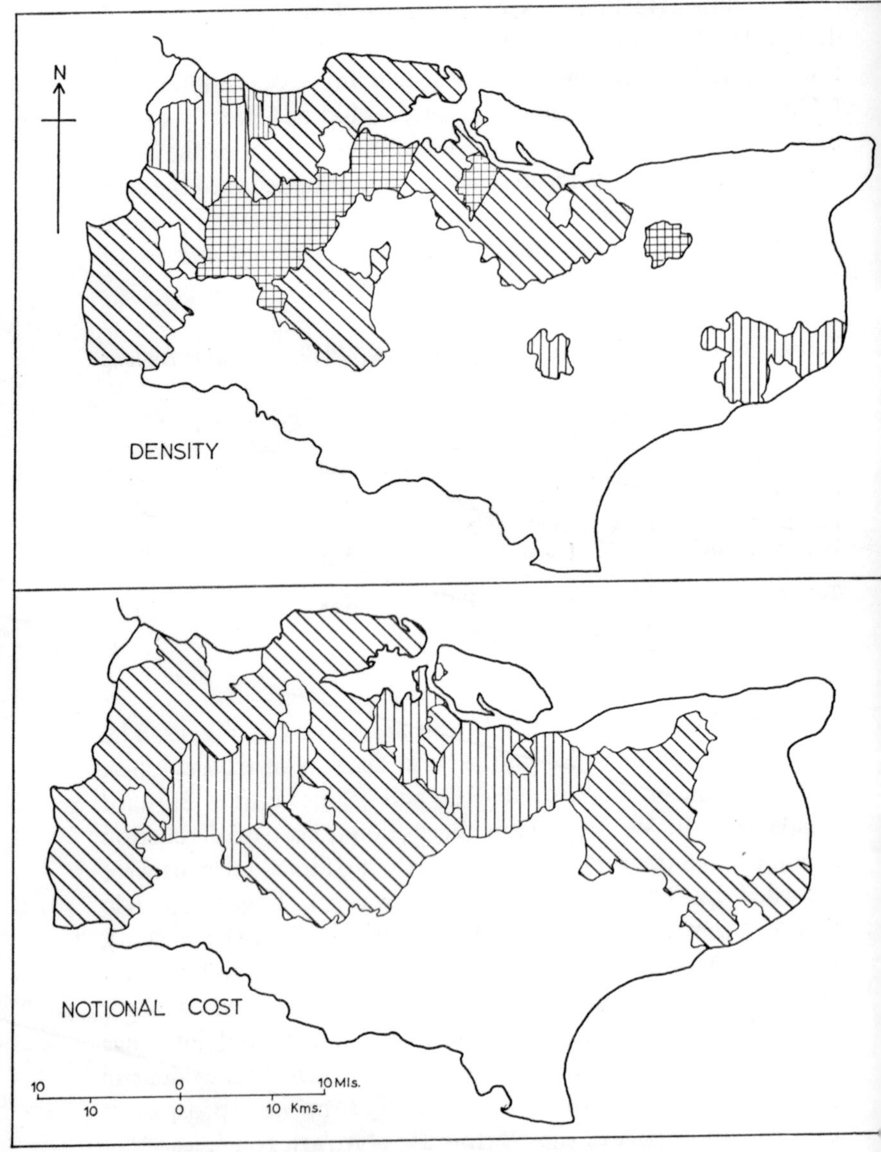

Fig 20 Derelict land in Kent, 1969 (see note below Fig 18)

Cornwall and Durham. Cheshire (see Chapter 9) is another example of the type of locality which lies between the high-density high-cost derelict areas and those districts with little or no dereliction, as depicted on Fig 14. In 1969–70 Kent stood out as one of the few localities in the South East with a density of derelict land above the national median value (1·1ha per 1,000). Even so local authorities controlling about 40 per cent of the county's area made nil returns, and roughly three-quarters of the recorded dereliction lay in the western third of the county. The main concentrations were on lower Thames-side between Gravesend and Sittingbourne, and along the Medway valley, notably in Malling RD, but even so very few localities recorded high densities of dereliction. Sittingbourne & Milton UD (11·54ha per 1,000), Canterbury CB (10·83ha per 1,000) and a group of three contiguous local authorities in the Medway valley (7·13ha per 1,000) recorded the highest densities, otherwise the values returned were at about the national median. Notional costs of dereliction were almost universally below the national median of 8 per cent and in no locality did they rise to the national upper quartile value of 21 per cent. Malling RD possessed the highest burden (14·5 per cent), but Kent in general exemplifies the prosperous county in which dereliction achieved no more than local significance as a land-use problem. The greater part of the derelict land comprised excavations and pits (67 per cent), mainly connected with the cement industry or with sand and gravel working. In addition there were small pockets of derelict land caused by coal mining (Coleman 1955), together with sections of abandoned railway and other miscellaneous small parcels of dereliction.

Leicestershire is one of a group of counties separating the major problem areas of northern and midland England from the areas of little dereliction further south (Fig 21). In 1969–70 its density of derelict land was 2·00ha per 1,000, just below the national upper quartile value, and the notional cost of dereliction was 6·5 per cent of total rateable value. The highest densities of derelict land were recorded on the Leicestershire coalfield, with 16·06ha per 1,000 in Ashby Woulds UD, and 12·23ha per 1,000 in Coalville UD. The first-named locality lies within the eastern part of the so-called South Derbyshire coalfield; the western part, centred on Swadlincote UD, had an even higher density (38·62ha per 1,000). Most of the derelict land stems from two causes, coal mining and clay working (Holmes 1958–60, Wallwork 1974), and includes abandoned spoil heaps, clay pits, and land damaged by mining subsidence. Although most of

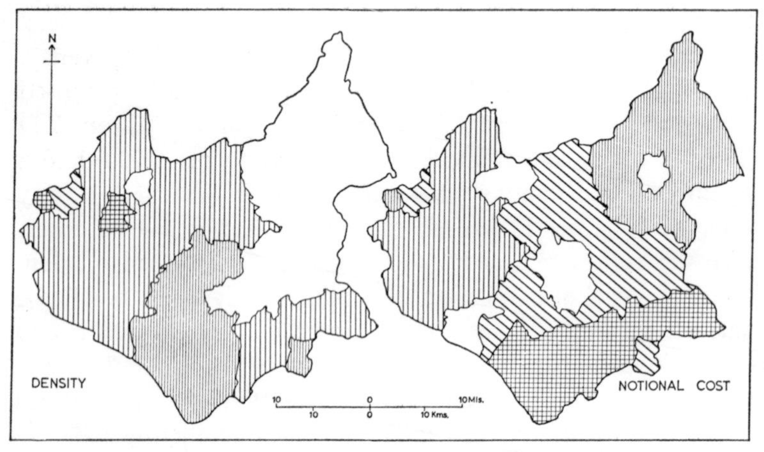

Fig 21 Derelict land in Leicestershire, 1969 (see note below Fig 18)

industrial Leicestershire 'nowhere . . . attains the dramatic and demented ugliness of the Potteries or the Black Country' (Hoskins 1957), this corner of the county, together with adjacent parts of Derbyshire, constitutes a small but singularly diverse problem area (Wallwork 1960). Elsewhere in Leicestershire abandoned 'granite' quarries and gravel workings form a major component of dereliction, notably on the margins of Charnwood Forest and along the Soar Valley. However, one of the largest sources of dereliction comprised abandoned railway land, notably in the eastern rural parts of the county, where several lines were closed during the 1950s. In 1964 derelict railway land accounted for 58 per cent of the county's total, and this propor-

Page 103 (*above*) Derelict land at Hedley Hope, 1967: low colliery spoil tips and other derelict land forming part of a 155ha site about to be worked opencast for coal; (*below*) The same site in 1972, after the removal of 252,000 tons of coal and restoration to agriculture

Page 104 (*above*) 'Hill and dale' produced by opencast ironstone working, Gretton Brook Road, Corby: photographed in 1954; (*below*) The same site in 1957 after restoration to agriculture

tion has remained throughout the 1960s. Although this form of dereliction does not produce high densities, it does constitute a challenging problem for local authorities in rural areas. The notional costs of dereliction were highest in several of the rural districts containing abandoned railway land, notably Lutterworth RD (47·7 per cent), Market Harborough RD (44·9 per cent) and Melton and Belvoir RD (23·7 per cent). High rate burdens were also recorded on the coalfield at Ashby Woulds UD (25·1 per cent) and Ashby de la Zouch RD (19·0 per cent). However, the county as a whole contained several highly rated local authority areas with little or no derelict land, and for this reason Leicestershire's notional costs were below the national median.

Finally Northamptonshire provides an example of a county in which the density of dereliction (0·62ha per 1,000) and its notional cost (3·3 per cent) are both below the national median. At first sight this is surprising in view of the county's long-standing importance as a source of iron ore and a centre of iron smelting (Beaver 1943). Roughly one-fifth of the county returned no derelict land, and this included many local authority areas in the iron mining districts. The highest recorded densities were in Irthlingborough RD (11·68ha per 1,000), comprising abandoned limestone workings at a former cement works, and Burton Latimer UD (5·40ha per 1,000) on the ironstone field. With the exception of 11ha of spoil heaps in Wellingborough UD, at the site of an abandoned ironworks, all the remaining dereliction in the county also comprised excavations and pits.

The development of mechanised mining on the ironstone field during the inter-war years posed a potentially serious threat of extensive dereliction through the creation of 'hill and dale', of which there were 566ha in Northamptonshire by 1951 (Civic Trust 1964). However, the application of the Ironstone Restoration Fund to this problem produced a rapid improvement and also minimised the long-term impact of mining after 1951 (see pp 242–3). The notional cost of dereliction is in general low, but both Irthlingborough UD (20·2 per cent) and Burton

Latimer UD (10·1 per cent) recorded high values by the standards of the county. More significant were the high values recorded in three rural districts, each with very low densities of dereliction: Kettering (20·6 per cent), Oundle and Thrapston (19·5 per cent) and Towcester (17·0 per cent). In Kettering RD the dereliction was caused by abandoned underground ironstone mines; in the other two districts most of the derelict land comprised abandoned ironstone and limestone quarries. Even in a prosperous low-cost county such as Northamptonshire, the rural districts bore a disproportionately high level of the notional costs of dereliction (Fig 22).

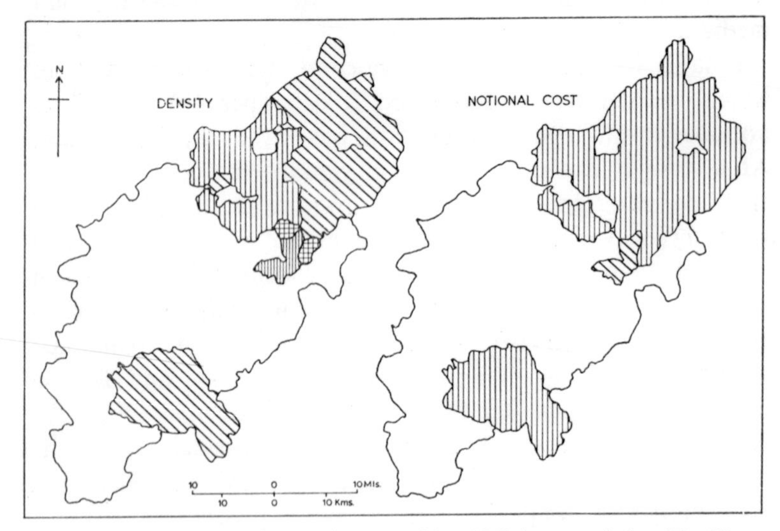

Fig 22 Derelict land in Northamptonshire, 1969 (see note below Fig 18)

CONCLUSION

Imperfect though they are in detail, the official returns of derelict land in England and Wales do provide a means of examining the distribution of dereliction at various spatial levels. However, even at the lowest level of areal refinement, the data relate to units which are both coarse and variable in size, so that any form of sophisticated analysis is rendered difficult irrespective of the reliability of the returns. The measurement of the cost of derelict land to the community is also problematical, not least because no consistent and universal estimates of real costs exist. The notional costs of dereliction hypothesised in this chapter have the virtue of being calculable for all levels of local authority area, but the assumptions upon which they are based are far removed from the current methods of financing the reclamation of derelict land. Nevertheless the notional costs of dereliction, expressed as a burden upon a local authority's rateable value, do provide a measure of spatial variations in the cost to the community of derelict land.

Five major points may be made in conclusion. Firstly, the distribution of derelict land conforms generally to expectations derived from a study of the historical development of mineral extraction and mineral-using industries, with two important exceptions. The opencast coal and ironstone mining industries have not produced dereliction, in spite of the great upheaval caused during extraction, because of statutory controls. The dereliction caused by abandonment of transport installations and industrial plant is less easily predicted than those forms of derelict land associated with mineral working, and does not, therefore, conform to expected patterns. Secondly, the density of derelict land varies appreciably at all spatial levels, and is perhaps most strikingly revealed in the analysis of local authority returns. Even at county level there is considerable deviation from the national mean, ranging from +15·42 in Cornwall to –3·00 in Hertfordshire, but with a mean deviation of only 2·82. Thirdly, dereliction is not wholly a problem of urban-industrial areas. Much derelict land lies in rural administrative areas, and although many rural districts contain substantial urban and suburban tracts (Robertson 1961), there are numerous rural areas proper with problems of dereliction. This is seen not so much in the density of derelict land, for the size of these administrative areas frequently masks the true extent of localised pockets of dereliction, but in the level of notional costs. The mean notional costs in local authority areas recording dereliction in the English counties with high densities of derelict land were 97 per cent for Rural Districts, 34 per cent for Urban Districts and Municipal Boroughs, and 5 per cent for County Boroughs. Fourthly, the notional costs of dereliction are themselves unevenly distributed, but not in a fixed relation to the density of derelict land. This is a corollary of disproportionately high burdens in rural areas, but it also applies at county level. Consequently, 'poor' counties with moderate densities of derelict land are in an apparently inferior position to 'rich' counties with heavy concentrations of dereliction. Finally the composition of derelict land also displays marked spatial variations, with a very wide dispersion of values at county level. The most significant feature here is the extent to which one of the three categories of derelict land dominates the composition of dereliction in many counties. The important generalisation to be made is that derelict land is unevenly distributed in terms of areal density, notional cost, and functional composition. This is partly a product of geology and the history of mineral exploitation, but it is also a reflection of social and economic variables, including variations in the pattern of demand for reclaimed land.

References to this chapter begin on p 310.

PART TWO

INDUSTRIAL CASE STUDIES

THE causes of dereliction can for convenience be divided into four groups:

1 mining and abstraction of minerals by pumping;
2 surface extraction of minerals;
3 mineral refining;
4 miscellaneous forms of industrial activity and transportation.

This fourfold division is based largely on the forms of derelict land produced by different modes of economic activity, or resulting from their cessation. Undoubtedly mineral working is a prime cause of dereliction, both in causing disruption during extraction and in producing solid residues during the processes of cleansing and refining. The abandonment of industrial plant, lines of communication and transport installations also produces dereliction, but it is of a less predictable form than that caused by most branches of mineral working.

Mining involves the abstraction of solids, including fossil fuels, metalliferous ores and, less commonly, non-metalliferous minerals, from artificially lit underground workings. Although the term is also commonly applied to surface workings (eg strip mine, opencast mine, *tagebau* – literally day mine), this narrower literal definition is employed in Chapter 4. Some minerals occur naturally as viscous liquids and are pumped to the surface, but as the forms of dereliction related to pumping are similar to those associated with mining, they are dealt with in the same chapter. Surface working is a self-explanatory term, but both the techniques employed and the minerals extracted are numerous: most of the dereliction stems from the fact that surface working creates excavations and waste tailings. The latter may in some instances resemble the residues produced by both mining and mineral refining, but their location close to quarries and pits places them in a different category when reclamation is envisaged. Consequently the forms of dereliction associated with surface working discussed in Chapter 5 include both excavations and waste heaps. In Chapter 6 several different forms of dereliction are reviewed. Mineral refining, together with some mineral-based industrial processes, may produce waste tips, some of which present more intractable problems than those associated with mineral extraction. Dereliction stemming from the abandonment of industrial plants and public utilities almost defies categorisation, and in consequence this will largely be examined in relation to the case studies of reclamation in Part Three. Derelict transport installations are more readily analysed in general terms, and they are considered in the final section of Chapter 6.

CHAPTER FOUR

Deep Mining and Dereliction

ALL forms of mining proceed by removing minerals from underground workings reached by shafts. The depth of the workings varies according to the stage reached by mining technology and the geology of the mineral being exploited. Similarly the type of shaft used and the disposition of surface installations reflect a combination of geological and technical influences. Two major consequences of mining may give rise to dereliction: firstly, the removal of the mineral may cause damage by subsidence of the overlying ground; and secondly, solid waste derived from the processes of mining and of mineral treatment at the pit head may be dumped on the surface.

COAL MINING AND DERELICTION

The extraction of coal is the most widely distributed cause of dereliction stemming from mining, for no other stratified mineral working yields such a combination of subsidence and waste tipping. Throughout the world these twin elements of coal mining have long attracted attention from engineers, lawyers, and land-use analysts, to say nothing of novelists from Dickens to Zola and D. H. Lawrence. In this account mining methods are considered solely in relation to the forms of dereliction they may cause. The two principal components of dereliction caused by coal mining and the attempts to mitigate their damaging consequences are discussed separately below.

Subsidence

Strictly speaking subsidence is the vertical displacement of the overlying surface which accompanies the removal of minerals from below ground. There is in addition an accompanying horizontal displacement, the mechanics of which are more complex, and this movement is responsible for the most serious disruption caused by mining, particularly in urban areas. In this account the term subsidence will be taken to cover both vertical and horizontal displacement, as illustrated in Fig 23a, in preference to the term 'ground movement', which is used by the National Coal Board (NCB) in Great Britain.

Subsidence caused by coal mining is probably the most widespread form of dereliction caused directly by mineral extraction below ground. Far from diminishing as techniques of mining have progressed, it has remained a persistent problem throughout the coalfields of the world. The form taken by vertical subsidence differs according to a wide range of technical and geological variables which may be summarised as follows:

1. thickness of seam;
2. depth and width of the area of coal extracted (in Great Britain termed 'goaf');
3. angle of draw; at a given thickness of seam and depth of working subsidence reaches its maximum extent when a certain width is extracted: this is expressed in terms of the angle of draw, which varies very widely from working to working;
4. the extent to which the 'goaf' is filled with solid waste after extraction has taken place;
5. variations in geological structure, notably faulting and folding of the coal seams, and differences in the tenacity of the overlying strata;
6. speed of extraction and method of underground working adopted.

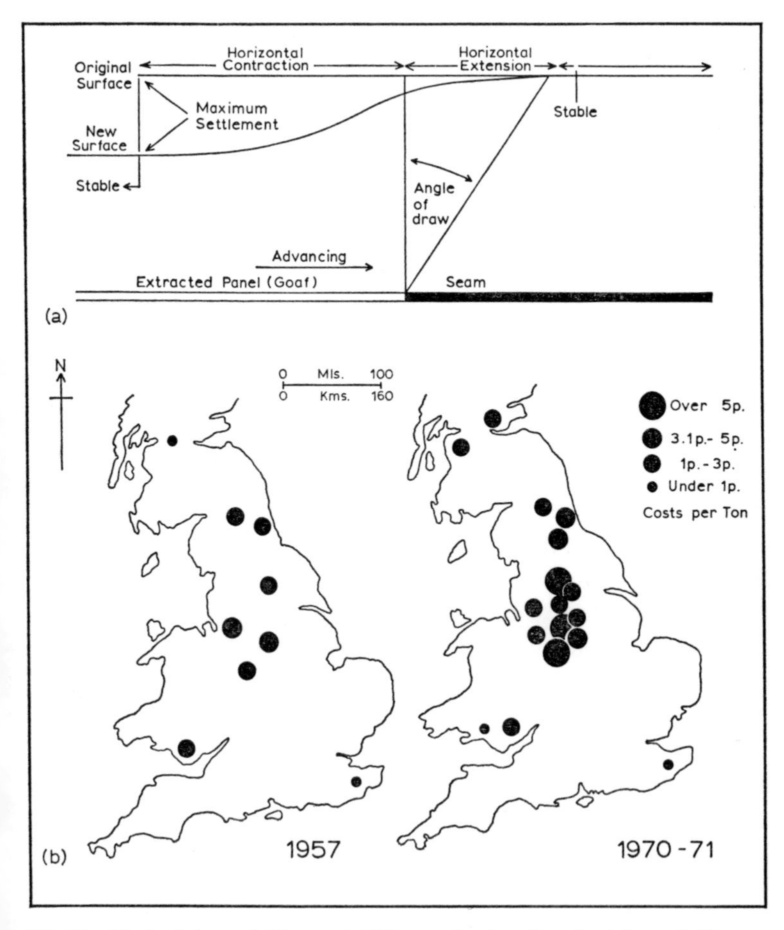

Fig 23 Coal-mining subsidence: (a) The mechanics of coal-mining subsidence; (b) compensation for surface damage, by NCB areas 1957 and 1971, in pence per ton of saleable coal by NCB divisions (1957) and areas (1971). Data from NCB *Annual Reports and Accounts*

Horizontal displacement is less predictable and accompanies vertical subsidence as a trough-like formation running parallel to the direction of coal extraction, progressing with a wave-like motion at the surface (Fig 23a). It is generally proportional to the amount of vertical subsidence and inversely proportional to the depth of mining.

As subsidence is partly a product of technique of extraction and partly a consequence of geological structure, it is necessary to outline the development of mining practice in relation to the forms of damage which stem from it. In most exposed coalfields primitive mining techniques were confined to small-scale working, either from shallow bell pits or from adits driven into the outcrop where the geology and topography permitted. Such workings may still be traced from the evidence of collapsed shafts and spoil mounds, but the damage caused was so slight and so highly localised that it is now scarcely more than an archaeological curiosity. The growth of deep mining, made possible in Great Britain from the mid-eighteenth century onwards by advances in the drainage and ventilation of mines and the advent of steam power, meant that coal pits became progressively larger, for their extent underground had to be sufficient to offset the capital costs of shaft sinking. Mines being both deeper and more extensive, the problems of supporting the roof of the workings became greater and by implication so did the amount of damage at the surface. Between 1801 and 1851 the maximum depth at which coal was mined in Great Britain increased from 270m to 650m.

The method of mining most commonly employed was 'pillar and stall' which, in its many local variants, involved partially working the coal to the boundary of the mining concession, leaving pillars of coal to support the roof at predetermined intervals. In many coalfields it was then the practice to mine the pillars, working back to the shaft and causing the abandoned galleries to cave in. Where very thick seams were being worked, it was possible to work the coal yet again once this collapse had compacted the debris containing workable coal from the mine roof. Thus although the pillar and stall method led to minimal subsidence during the initial phase of mining, substantial damage commonly took place when the supporting pillars were removed, and even where they were left *in situ* the strains and stresses imposed by the overlying strata might cause more gradual subsidence to occur after the mine had been abandoned.

Similarly, subsequent mining at greater depth beneath the abandoned pillar and stall workings has been a common cause of unexpected rapid subsidence, particularly when the presence of the earlier workings was unsuspected.

Under British mining conditions the continued use of pillar and stall extraction was not well suited to the greater depths of working of the second half of the nineteenth century, and in several coalfields the technique was steadily abandoned. It was replaced by various forms of 'longwall' mining, except in those localities which retained pillar and stall for a variety of technical and traditional reasons. By 1948 only 12 per cent of British coal production came from pillar and stall workings, and in 1972 the proportion had fallen to 5·5 per cent, almost equally divided between mechanised and manual extraction. The only NCB area producing the bulk of its coal by this method was Northumberland, with 40 per cent of its output from mechanised pillar and stall workings and 21 per cent hand got.

Longwall mining exists in two principal forms: longwall retreating, which works from the outer limit of the mining concession back to the shaft or a main haulage road, and longwall advancing, which proceeds out from the shaft bottom to the limits of the mining concession. In 1972 10 per cent of British coal came from longwall retreating faces, compared with 83 per cent from longwall advancing. Almost all the longwall advancing faces were fully mechanised, and this method of extraction was particularly important in the highly productive fields of South Yorkshire and the East Midlands. The major effect of longwall mining is that normally all the coal is removed as the face advances, leaving voids equal to or even greater than seam thickness. This last situation occurs where thin seams are being mechanically worked and it is necessary to mine barren material in order to accommodate the machinery. Some support is provided by loosely packed waste and the lining of haulage roadways in the goaf, but this is largely to facilitate working and does very little to minimise subsidence. As mining proceeds, a trough of subsidence is created at the surface.

The detailed mechanics of its formation are dealt with in several technical publications (Institution of Civil Engineers 1959, National Coal Board 1965) and little need be said of them in this account. Two points should, however, be made. Firstly, vertical displacement by subsidence is normally less than the thickness of the seam being worked, and becomes slighter with an increase in the depth of mining. This is because the collapse of the overlying strata causes them to fragment and fracture and the mass of rock fills a greater space than it did when naturally compacted. Consequently subsidence at the surface may be equal to no more than 30 per cent of the thickness of the coal removed from deep workings. Secondly, the most costly damage caused by subsidence, that affecting buildings and services, is a product of horizontal displacement, which, although commonly disruptive in urban areas, rarely gives rise to permanent dereliction. At its worst buildings may be structurally damaged to the point where demolition is necessary for reasons of safety, but once mining has ceased, the site can readily be put to economic use again. Admittedly this type of damage is often most apparent in rows of terraced houses, which are particularly vulnerable to subsidence and which may also merit demolition because of other substandard qualities. However, the apparently deserted sites left by the demolition of such properties are not normally included as derelict land, just as the equivalent sites of clearance areas in subsidence-free localities are not so classified.

The relationship between subsidence and derelict land is therefore largely confined to two sets of circumstances. Firstly, subsidence may cause the disruption of surface drainage and the resultant depressions then become permanently flooded. Extensive subsidence lakes are most likely to form in low-lying areas which are naturally liable to seasonal flooding. In England the Anker valley in East Warwickshire (Wallwork 1960b), the valleys of the Aire and Calder in the Castleford-Knottingley area (Collins and Bush 1971), and the Wigan-Leigh area of Lancashire (see pp 266–70) provide examples of subsidence lakes formed on the floodplains of streams. Although they represent a substantial loss of land, these bodies of inland water may, with careful planning, be turned to beneficial use at relatively little expense when schemes of reclamation are being undertaken. Secondly, subsidence may produce an irregular pock-marked surface which, even if water does not permanently collect in the hollows, cannot easily be put to economic use. Much of the dereliction in the South Staffordshire and Shropshire coalfields was of this kind, comprising numerous collapsed shallow mines interspersed with low hummocks of colliery spoil (see pp 279–83).

Not surprisingly coal-mining subsidence has frequently been the subject of parliamentary enquiry in Great Britain, although major legislation dealing with the problem dates only from 1950. Interest in mining subsidence has almost entirely centred on the question of financial liability for damage caused, rather than on removal of the source of the damage itself. In 1927 a Royal Commission enquired into the possibilities of providing compensation for subsidence without producing a workable solution and, as the pages of *Hansard* show, successive governments were unable to unravel the complexities of this issue. A second governmental enquiry in 1949 was more successful, for it gave rise to the Coal Mining (Subsidence) Acts of 1950 and 1957.

The legal niceties of compensation for damage caused by coal-mining subsidence are not the concern of this account, but the consequences of the complex legal position and the relation between dereliction and legislative measures need to be outlined. In simple terms no colliery company had an obligation to compensate for the damage normally caused by subsidence before the passage of the 1950 Act unless the mining lease required it in relation to land overlying the mine workings generally, or, more commonly, to specific surface structures. Many colliery companies ensured that all possible liabilities were avoided by buying the land most likely to be damaged, using some of it for waste tips and other ancillary functions, and leasing the remainder to tenants without any compensatory rights. Thus in 1946, when coal mining was nationalised, the NCB owned approximately 141,000ha of land (excluding that occupied by colliery housing), of which less than 0·5 percent was derelict. Some colliery leases included clauses which were intended to mitigate the ill effects of mining at the surface, but the practice appears to have become uncommon during the second half of the nineteenth century, when most of the major pit sinkings took place. Thus a colliery lease at Brownhills, Staffordshire, in 1870 permitted extraction to take place regardless of damage caused at the surface, and implied that this was a widespread practice. Various parliamentary enquiries into the coal trade and the working conditions of miners also noted, *en passant*, that the practice of leaving pillars of support, in order to avoid claims for damage at the surface where litigation was possible, was uneconomic, as the cost of damage in monetary terms might be trivial in relation to the loss of unworked coal. As late as the 1930s the parliamentary attitude to local authorities who complained of the disruption caused by coal mining was that they should consider the provision of employment and rateable income by colliery companies as more than adequate recompense for the damage caused by subsidence.

The post-war Acts placed responsibility for compensation upon the NCB, which, in return for this obligation, was empowered to lay down conditions governing the construction of new buildings in subsidence-prone localities. One consequence of the Acts was, therefore, a reduction in the volume of building in localities threatened by subsidence in order to avoid the extra cost of meeting the stringent requirements of the NCB. In addition local planning authorities are obliged to work in consultation with the NCB in order to phase development with the various stages of coal extraction, thereby minimising the risks of damage to buildings and services. Nevertheless compensation for damage caused by coal-mining subsidence continues to be paid at a rate of approximately 4p per ton of saleable coal produced (Table 7).

Page 121 (*above*) Gravel pit restored to agriculture, Great Waltham, Chelmsford. The site forms part of a 68ha scheme of gravel working and phased reclamation, with pits worked to a depth of 16m; (*centre*) Gravel pit restored by solid filling and used as a site for housing, Rainham, Essex; (*below*) Amenity use of a 40ha former wet gravel pit, Dorchester on Thames, Oxfordshire

Page 122 (*above*) Derelict land at Bamfurlong colliery, near Wigan, 1962, showing weathered colliery waste; (*below*) The same view in 1970 after reclamation for agricultural use at a cost of £750 per hectare. This locality appears in plate on p 140 and is depicted in Figs 35a and 35b

TABLE 7 Compensation for surface damage (million pounds sterling)

Year	£m	Year	£m	Year	£m	Year	£m
1947	0·6	1953	2·6	1959	3·7	1965	4·6
1948	0·9	1954	2·6	1960	3·4	1966	5·0
1949	1·2	1955	3·1	1961	3·2	1967	5·2
1950	1·4	1956	3·5	1962	3·8	1968	4·9
1951	1·8	1957	4·3	1963	4·9	1969	4·6
1952	2·3	1958	4·3	1964	3·9	1970	5·5
						1971	5·7

Payments during the period 1947–56 were made under the provisions of the 1950 Coal Mining Subsidence Act; from 1957 onwards the Act of that year has applied

Data from NCB *Annual Accounts* 1957–71; *Hansard*, vol 617, Session 1959–60, p 2

In 1957 the national average rate of compensation for surface damage was 2p per ton of saleable coal. There were, however, substantial local variations, and the median value was only 1p per ton. The NCB areas with low rates of compensation covered the Scottish coalfields and Northumberland, together with parts of the Yorkshire, Nottinghamshire and South Wales coalfields (Fig 23b). The NCB areas with high rates of compensation (above the upper quartile value of 2·5p per ton of saleable coal) were widely scattered throughout the English and Welsh coalfields but lay most notably in Durham, Lancashire, Derbyshire and the Swansea valley. Almost invariably the highest rates were in urban areas, notably Manchester (8p per ton), Ilkeston (9p) and South Derbyshire (7p). The data for 1971, the last published, were based on a different set of areas, but the broad pattern of variation in rates of compensation was much the same (Fig 23b). Durham and Derbyshire contained areas with high rates of compensation; Scotland and Wales remained as areas with exceptionally low rates. The major difference lay in the transference of the exposed coalfield of Yorkshire from the low-cost to the exceptionally high-cost category. In 1971 the national rate of compensation had risen to 4p per ton of saleable coal. This was not necessarily a reflection of greater damage, for inflation, coupled with a diminution of the tonnage of coal mined, might account for a substantial part of the increase.

Even so, the rate of compensation is undoubtedly much less than the cost of preventing excessive damage by employing various preventive measures such as underground stowage of waste. In many respects the 1950 and 1957 Acts set the seal of respectability on coal-mining subsidence by providing the right to compensation in exchange for the obligation to erect new buildings in areas of risk to very stringent standards of construction, instead of minimising the damage by taking preventive measures below ground.

Two alternatives exist if the provisions of the 1950 and 1957 Acts cannot be met by proposed forms of development, including the reclamation of derelict land. Firstly, development may be postponed until 10 years after mining has ceased and the risk of subsidence has passed. Examples of this type of deferment are commonplace, one being the newly built housing estates on the abandoned coalfield south of Wigan (see plate, p 140), which partly occupy formerly unstable derelict land. Secondly, the coal may be left unworked or only partially extracted where the problems of damage at the surface are particularly severe. For the most part this kind of provision has been limited to carefully selected urban areas and cannot be considered as a measure to prevent dereliction. However, there is an outstanding instance in which partial extraction was adopted as long ago as 1929 in order to prevent widespread flooding of high quality farmland. The Doncaster Area Drainage Act (1929) was the only positive action to stem from the findings of the 1927 Royal Commission. This Act provided for surface drainage works to be carried out by the colliery companies, thus enabling one or two seams to be worked without the risk of permanent flooding. No coal was to be worked from beneath the River Don, owing to the great cost of raising its banks to compensate for the subsidence that would inevitably occur.

Colliery waste disposal

Subsidence is an unavoidable element of coal mining capable of mitigation, but it can be argued that the dumping of waste at the surface ought to be avoided with much less difficulty. Colliery waste has two components: the solid rock brought to the surface during shaft sinking and the driving of roadways though barren strata, and the solid residues produced by coal preparation plants at the pit head. The first source of waste is normally insignificant after the initial shaft sinking, when it may yield as much as 150,000m^3 of rock per 1000m of double shaft. The use of horizon mining, formerly employed in localities with steeply dipping strata, also created rock waste, as the technique involved driving horizontal headings through barren ground in order to intersect the productive coal seams. However, most colliery waste comes from the cleansing plants which separate saleable coal from the waste residues.

Mechanised mining has greatly added to the volume of waste because it is less selective than manual working in rejecting non-combustible material, which used not to be brought to the surface. The trend towards fully mechanised mining in Great Britain has, therefore, added to the volume of waste brought to the pit head at a time when total output of coal has been falling. In 1919 approximately 12 per cent of British coal was mechanically cut; the proportion rose to 72 per cent in 1944 and 92 per cent in 1960. During the 1950s fully mechanised mining was introduced, coal being both cut and loaded on to conveyor belts by power for transmission to the shaft. In 1972 92 per cent of British coal came from fully mechanised faces, compared with 38 per cent in 1960 and 4 per cent in 1950, and this has again added to the amount of waste brought to the surface. In addition the proportion of the coal mechanically cleaned at the surface has also greatly increased, from 28 per cent of total output in 1920 to 50 per cent in 1948 and 68 per cent in 1967. Consequently the ratio of waste to saleable coal has also increased since the 1940s. In 1948 the proportion of waste to saleable coal was 20 per cent; by the 1960s it had risen to 25 per cent, and recent estimates (Denison 1969) suggest that 30–35 per cent of the tonnage raised at collieries is waste.

Like subsidence, waste disposal did not constitute a major

problem in coal-mining areas during the first half of the nineteenth century, for as long as shafts were relatively shallow the coal was worked manually, and as the techniques of cleaning remained primitive, very little waste required tipping at the surface. Such tipping as occurred was not mechanised, so that waste heaps were both small and low. The progressive deepening of shafts and enlargement of individual pits added to the volume of waste raised to the surface at a particular point, but even more important was the increased use of mechanised mining and coal preparation as a source of spoil. Initially the larger tips required at modern collieries were created by tipping from tramway tubs and railway wagons, producing gently graded spoil heaps which threatened to become excessively space-consuming as they fanned out from the original dumping ground. The introduction of fully mechanised tipping gear appeared to solve this problem by allowing spoil heaps to rise to heights of 50m or more. The mechanically produced conical and ridge tips (the precise form of which depended on the type of tipping gear used) became common features of the coalfield landscapes of Great Britain during the first half of the twentieth century as new areas of deep mining from large collieries were opened up. Technically these tips were not as satisfactory as had been assumed, for lack of compaction in the tipping process left 30 per cent of their volume as air voids, and this in turn made them prone to spontaneous combustion and instability.

Ironically it was the problem of spontaneous combustion which first aroused governmental concern about the surface disposal of colliery waste, after much prodding by members of Parliament representing mining constituencies. Even so the legislation framed in 1939 was designed not to eliminate surface tipping but to reduce the number of blazing spoil heaps, which had incidentally become a threat to military security in their fortuitous role as beacons guiding possible enemy air attacks. The 1939 Act placed an obligation on colliery companies to attempt to extinguish or otherwise control burning spoil heaps, but it was not until the passage of the Clean Air Act of 1956 that the NCB was obliged by statute to take every reasonable step to combat spontaneous combustion as opposed to merely controlling the flames. Since 1956 hardly any new conical or ridge tips have been formed, although there is a legacy of older tips of these types. Again the 1956 legislation was designed to control surface tipping rather than to eliminate it, and the same can be said of the most recent legislation, the Mines and Quarries (Tips) Act of 1969. This was passed as a response to the Aberfan disaster of 1966 and is designed to ensure that spoil heaps do not constitute a hazard to life and property. The principal consequence of the 1956 Clean Air Act (as it applies to spoil heaps), reinforced by the safety provisions of the 1969 Act has been to substitute wagon tipping for other mechanical methods of waste disposal.

Wagon tips are normally created by rubber-tyred dump trucks which can most easily operate on gently sloping or level surfaces and in doing so compact the waste previously tipped. Consequently most of the recently created colliery spoil heaps comprise plateau-like surfaces which rarely rise to a height of more than 15m in all, thereby posing fewer problems of reclamation. Indeed tips of this kind can be reclaimed as dumping proceeds on other parts of the site, which was not possible with conical and ridge tips. The fact that modern spoil heaps can be reclaimed shortly after they have been created is evidence of progress towards solving the problem but not of completely eradicating it in the coalfields.

Three alternative modes of disposal which avoid the creation of waste heaps appear to exist, other than tipping spoil over the cliffs and into the sea as practised in County Durham. The first would involve a controlled dumping of waste at sea, using hopper barges similar to those which take towns' refuse to be tipped off the approaches to the Thames estuary. Although this would be technically feasible, the capital costs and organisational problems, coupled with the possibility that the waste might ultimately be borne back to the coast by tidal currents, render it impractical. The second mode of disposal – stowage of waste in disused mine workings – has long appeared to be practicable.

The report of the Chief Inspector of Mines in 1919 observed that 40 per cent of the coal underlying populous districts was left as support, and considered that hydraulic stowage of waste or other solids would protect the surface against risks of subsidence while permitting total extraction of the coal, as was then the practice in parts of Japan. In 1931 one of the attempts to introduce legislation designed to secure compensation for damage from subsidence mentioned the possibility of introducing back stowage of waste, as was done in several German collieries. In both instances it should be noted that the primary aim was not to reduce the volume of waste tipped at the surface, although this would have been an incidental benefit.

With the adoption of longwall mining, waste began to be used for packing the goaf as the face workings progressed and it was necessary to maintain haulage ways between the face and the shaft. However, this material is only loosely packed, and it is necessary to provide a much greater degree of compaction if the amount of subsidence is to be reduced. The technique of pneumatic stowage achieves this end, and is the principal method used in British coalfields. In 1946 the NCB operated six pneumatic stowing machines; 10 years later sixty-eight were in operation, capable of stowing roughly 2,000,000m^3 of waste per annum. In 1972 356 pneumatic stowing machines had been installed at British collieries, mainly in Yorkshire (135 machines), Northumberland-Durham (91 machines), and Nottinghamshire–Derbyshire (80 machines).

In spite of this post-war increase in the number of machines installed, it has frequently been argued by the coal industry that the cost of back stowage is unduly high, even in heavily urbanised areas where shortage of space for surface tipping and the benefits of reduced claims for damage by subsidence might be thought to exert powerful influences in its favour. Thus in 1945 a report on the coalfields of the Midland region observed that 'experiments with solid stowing on a large scale have been undertaken in North Staffordshire (but) the cost . . . is altogether prohibitive'. On several occasions the annual reports of the NCB have reiterated this point, noting that it would be necessary to devise cheaper methods of underground waste disposal if the practice were to become widespread. Much of the back stowage is not designed solely to reduce the volume of waste tipped at the surface. Often the technique is employed to protect reserves of coal that might not otherwise be worked. Thus in 1949 the NCB noted that 'by reducing the danger of spontaneous combustion and the amount of subsidence, power stowing can make it possible to work reserves which would otherwise be abandoned. But it cannot be used more widely unless it can be made much cheaper'. An outstanding example of underground stowage took place at Donisthorpe Colliery, Leicestershire, in 1956, when waste was conveyed underground from the surface via specially drilled boreholes at a rate of 1·4 tons of waste for every ton of coal mined. But this too was to protect reserves of coal, not to diminish dereliction. It is in fact argued that the speed of modern mechanised mining is so great that the return of waste to the workings would impede production of coal while doing little to reduce subsidence, which has, in any case, become a less serious problem in areas of high-speed deep mining.

Underground stowage of waste is, therefore, technically feasible in many collieries, and the most persistent arguments against its universal adoption have been economic, based on the increased capital cost of installing equipment to handle underground waste disposal and, with the advent of high-speed mining, the operational costs that would result from working at less than full efficiency in order to accommodate back stowage. However, the financial case against back stowage as an alternative to surface waste disposal has never been fully substantiated by the NCB. Generally speaking the cost of surface tipping is based on the expense of providing, operating and maintaining the tipping equipment. The cost of land is not normally included, for much of it has long been in the hands of the coal industry, and the expense of eventual reclamation as a charge against tipping costs is ignored. It would require very much more sophisticated cost-benefit analysis than the available data permit to know whether back stowage was much more expensive than surface disposal of waste. As it is, back stowage is most likely to occur when justified by mining needs, or, less commonly, where damage to overlying urban areas can be minimised by solid stowing. It is most unlikely to be employed solely as a means of environmental improvement, either in removing existing tips underground or in preventing the creation of new ones.

The third alternative to the creation of waste heaps is to find a market for the solid material contained in the tips. For many years red shale extracted from burnt spoil heaps has been accepted as filling material for various types of civil engineering work, but in Great Britain the use of raw shale has only been permitted since 1968–9. As modern spoil tips do not burn, this decision to allow the use of black shale provided an important outlet for the waste currently being produced at collieries, in addition to providing an incentive to clear older black shale tips for which no ready market had previously existed. The NCB markets shale of both types at a rate of approximately 7 million tons per annum, or about 15 per cent of current waste yields. Much of the shale is used in road construction as a filling material, often as an alternative to excavating borrow pits along the line of new highways. In addition colliery waste has been used as fill at other types of civil engineering project, including the Llanwern iron and steel plant in Monmouthshire and the Pegwell Bay hoverport in Kent. More ambitious schemes have been proposed, including the shipment of spoil from North East England to Holland for land reclamation and to Maplin airport as filling material. In both instances the ability to secure cheap bulk transport for the shale will be a crucial factor in determining the feasibility of the projects. Lack of such facilities prevented colliery waste being brought from South Wales to be used as filling material in the construction of the London–Bristol motorway. Even if major sources of consumption of waste do emerge, they will still fail to dispose of all the spoil brought to the surface in the foreseeable future, to say nothing of failing to make an appreciable impact on existing colliery tips.

It is also claimed that colliery spoil can yield valuable economic minerals such as potash and alumina, although it remains to be seen whether laboratory experiments proving this to be so can be translated into full-scale manufacturing plants. It is also the practice to rework some coal tips for small coal which was discarded during earlier periods of mining, but this again does little more than nibble at the problem, as production from tips and slurry ponds between 1951 and 1970 totalled about 23,000,000 tons. In contrast colliery spoil is being tipped at a rate of approximately 45,000,000 tons per annum and additionally consumes roughly 300ha of land each year. Fortunately the removal of existing spoil heaps is not entirely conditioned by economic criteria but, notwithstanding the existence of major schemes of reclamation (see pp 235–8), the spoil tip is likely to remain a feature of the coal-mining landscape for many years to come.

Coal mining in Great Britain

Although modern methods of mining have tended to reduce the amount of serious damage caused by subsidence, and parallel developments in the disposal of colliery waste, whether below ground or at the surface, give promise of less intractable problems of reclamation, the fact remains that coal mining has left a considerable legacy of dereliction. The British statistics of derelict land do not identify the causes of dereliction, but it is possible to estimate part of the coal industry's contribution to the total amount of dereliction by examining the returns made by local authorities. In 1971 there were 14,300ha of spoil heaps in England according to the official record; of this area roughly 56 per cent may be attributed to coal mining. In addition an incalculable proportion of the land classified as 'other forms of dereliction' will include such elements as abandoned colliery buildings, disused mineral lines, and land permanently flooded through coal-mining subsidence.

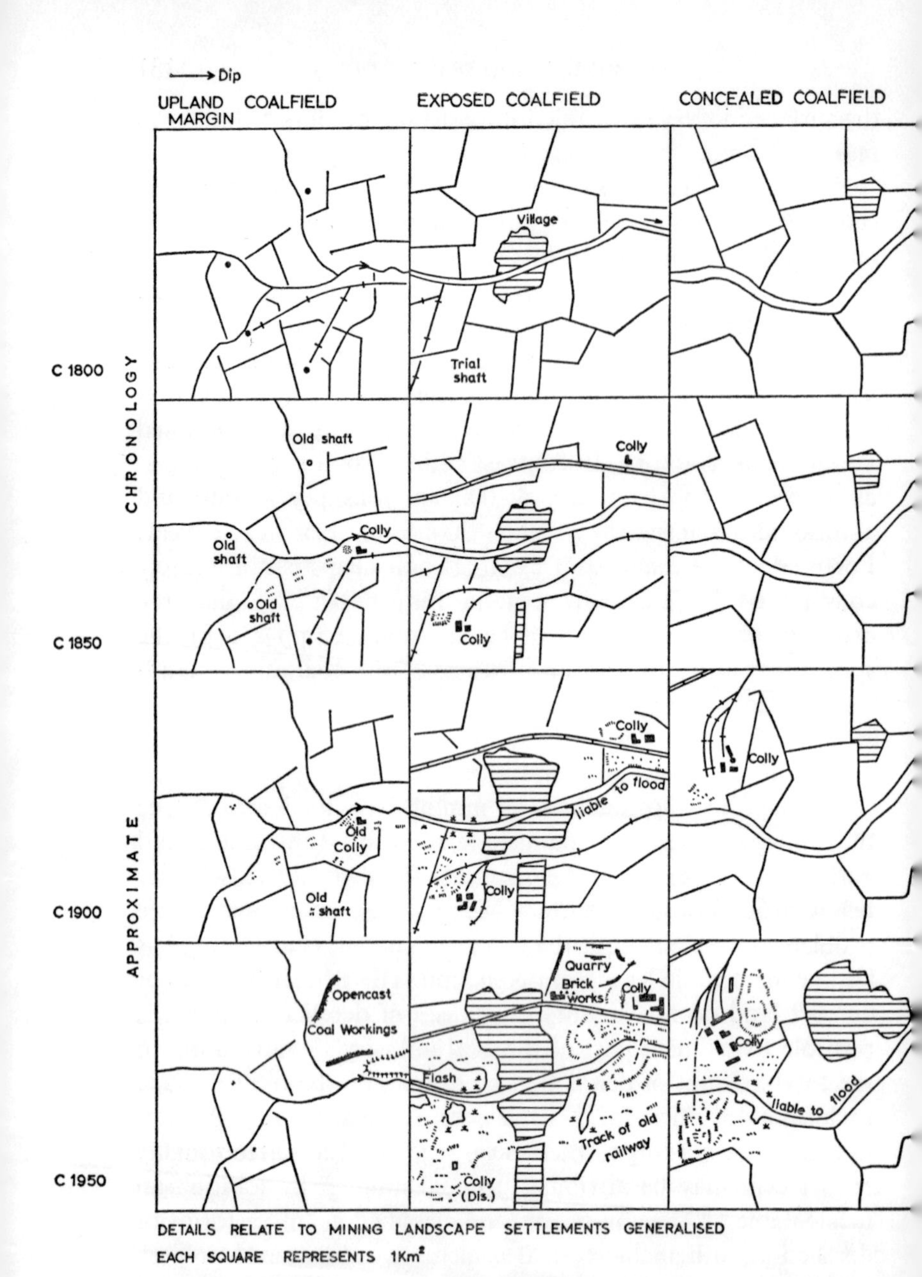

Fig 24 A model to illustrate the development of mining and dereliction in a British coalfield. It is assumed that the coal seams dip uniformly from left to right. The dates of development indicate the general timespan involved

The stages of development of dereliction in a mining locality are illustrated in Fig 24, which takes the simple case of a coal-field in which the strata are dipping uniformly without any structural disturbance. In a symmetrical syncline the same sequence of events might be expected to occur, working from the outcrop to the centre of the trough from both extremities. The importance of technical advances in mining is clearly demonstrated by the changing scale of dereliction. On the outcrop small-scale workings during the earliest phases of mining gave rise to a low amplitude of dereliction, characterised by small manually tipped spoil heaps and highly localised subsidence. Much of this dereliction was subsequently obscured by vegetative cover and then eradicated by opencast reworking of the outcrop from the 1940s onward. With the shift of mining down dip, and the consequent enlargement of collieries mining at increasingly greater depths, the scale and amplitude of dereliction also became greater. Subsidence became more commonplace, particularly in localities susceptible to natural flooding, where permanent lakes formed. Spoil heaps began to grow in volume as the amount of waste from below ground and from the washeries increased, and in height as various mechanical means of tipping were introduced. In recent years the problems of dereliction have been compounded by the accelerated rate of colliery closure, producing a mixture of abandoned pit-head buildings and ancillary plant, disused mineral sidings and railway tracks, in addition to the almost inevitable spoil heap fringed by other forms of waste land.

In relation to British coal-mining practice three general points may be made in conclusion. Firstly, the creation of derelict land in the coalfields of Great Britain seems likely to persist during the immediate future and, even though the rate of dereliction may be abated, the backlog of damage done in the past is very great. Secondly, the most severe problems of dereliction, and the most costly ones to redeem, are not a product of the early phases of mining (even though they may frequently be described as 'awful legacies of the Industrial

Revolution' in popular accounts) but of the recent past, when new techniques of extraction and coal preparation were pressed forward at a much more rapid rate than those devised to reclaim land or to mitigate the worst effects of mining. Finally, there is growing evidence that the equation of coal mining with large-scale disruption of surface land use is no longer accepted as inevitable, even within the coal-mining industry. This has never been the case with modern opencast coal working in Great Britain (see pp 165–6); somewhat belatedly and hesitantly the deep mining sector of the industry is moving towards a similarly enlightened attitude. Much depends, however, on the place accorded to coal in future decisions on national fuel policy and the extent to which any statutory obligations to minimise the extent of new dereliction are financially underwritten by the state or by the consumer.

Coal mining abroad

Dereliction in coal-mining areas is not confined to Great Britain, although there appears to be a widely held belief that techniques such as underground waste stowage are more commonly employed abroad, particularly in Western Europe. For example J. R. Oxenham observed in 1966 that 'stowage has been little used in this country (Great Britain), although it has been widely adopted on the continent'. While it is true that underground waste stowage has been practised in the more heavily urbanised parts of the Ruhr for over 60 years, within that coalfield there is still ample evidence of the effects of coal mining at the surface. This is so even in the most recently developed sectors of the coalfield: for example, at Marl, in the Emscher-Lippe zone, the Auguste Viktoria colliery has produced 105ha of spoil tips, including a 750m long ridge tip fed by an aerial ropeway, and this is by no means an isolated example (SVR 1960). In addition lakes and marshland caused by mining subsidence are to be found in many parts of the Ruhr (Wohlrab 1967). Bottrop, a mining town of the Emscher zone with collieries dating mainly from the early years of the twentieth century, had problems of waste disposal and mining subsidence in the 1950s, in spite of the fact that the most modern of its pits stowed waste below ground (Vogel 1959). Perhaps the most significant feature of mining practice in the Ruhr has been the extent to which spoil heaps have been reclaimed and afforested, rather than their having been completely avoided by the widespread adoption of underground stowage, as is sometimes implied. The programme of 'turning the Ruhr green', initiated by the regional planning authority in 1926 (see pp 237–8), was largely a reflection of the extent of dereliction in the locality, much of it a product of coal mining.

In other European coalfields the twin elements of subsidence and surface disposal of waste are also to be found. In northern France conical spoil heaps (*terrils*) are interspersed among the mining settlements (Fig 25) and, although about 20 per cent of the colliery waste is returned underground, subsidence is a great problem, particularly in low-lying areas. In Belgium similar forms of tip dominate the plateaux of Herve and Hesbaye near Liége, where 'great spoil banks, the product of over 100 years of modern mining, overlook the small houses of the miners' (Elkins 1956). Conical spoil tips are also a feature of the heathland scene in the much younger Kempen coalfield, developed during the 1920s, emphasising the point that the continental coalfields were not necessarily more advanced than those of Great Britain in their mode of waste disposal.

One of the most interesting mining practices, which has on occasion been suggested as a means of reducing the problem of subsidence in British coalfields, is that found in Silesia (Pounds 1958). It was introduced to the Polish sector of the coalfield in the early years of the twentieth century, largely to provide roof support in mines where coal between 40m and 60m thick was being worked within 1,000m of the surface. Under these conditions traditional means of supporting the roof were unsuitable and, in the absence of adequate amounts of local mine waste, sand and gravel were piped into the workings. The survival of the *podsadzka* method enables a higher proportion

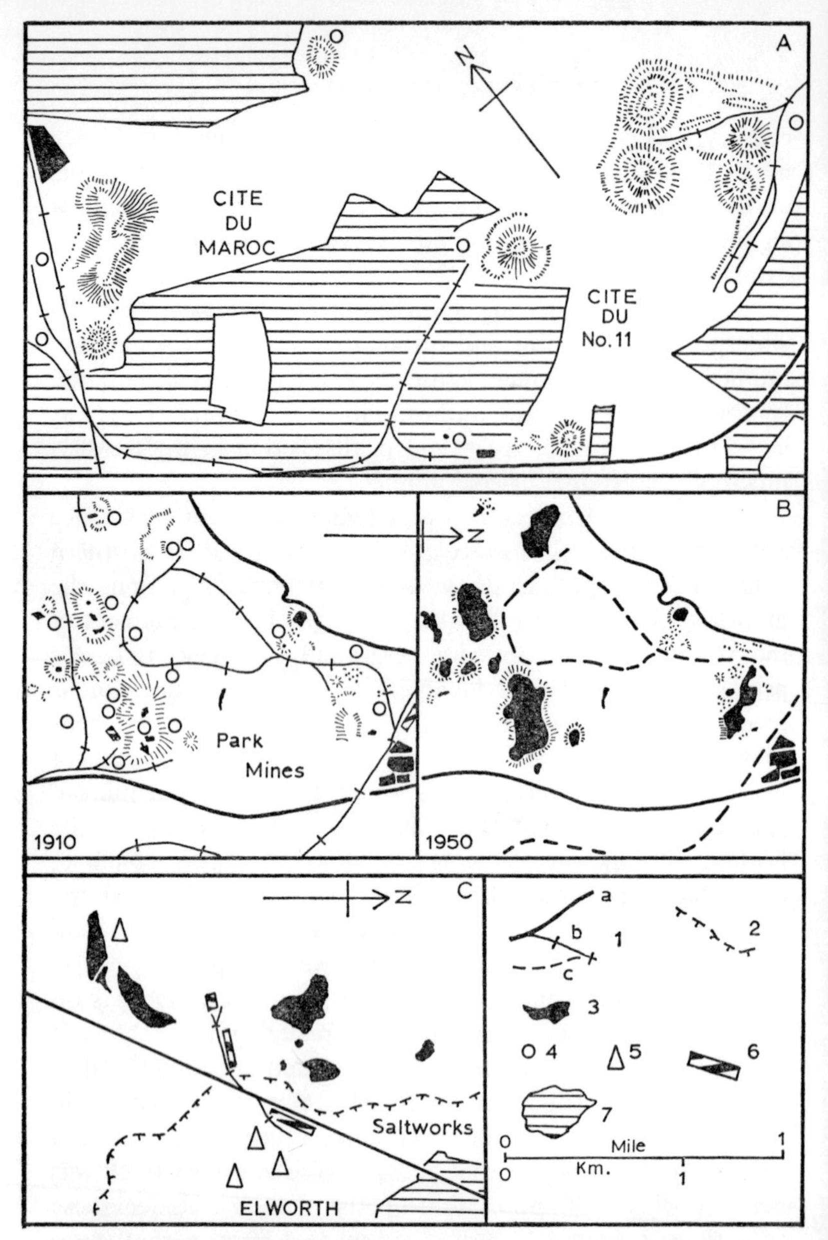

Fig 25 Landscapes of deep-mining areas

A The Nord coalfield north-west of Lens, showing the inter-mixture of colliery spoil heaps (*terrils*) and settlements

B The haematite iron-mining district north of Barrow-in-Furness at two dates: 1910, when active; and 1950, when derelict

C The Cheshire saltfield near Sandbach in the 1950s showing subsidence flashes caused by brine pumping

1a, railways; b, mineral lines; c, disused mineral lines; 2, canal; 3, water; 4, mine shaft; 5, brine pump; 6, industrial plants (generalised); 7, settlements. All the maps are to the same scale

of the coal to be extracted and permits workings to extend under urbanised areas with a minimal risk of subsidence. Beneficial though this may be to the coal-mining industry, the working of the sand and gravel itself poses problems of land use, in addition to adding to the pit-head price of coal. Even so, large tracts of agricultural land have been damaged by subsidence in the Silesian coalfield (Klimaszewski 1961), with the formation of trough-like depressions up to 10m deep. In Poland, as in Great Britain, the most expensive methods of underground support are only introduced beneath the major urban-industrial areas, and then as much to improve the yield of coal as to reduce the impact of dereliction. In the USA colliery waste has commonly been tipped at the surface, and there have also been considerable problems of subsidence, particularly in Pennsylvania (Grey and Meyers 1970) where experiments with underground stowage of waste were being carried out during the early 1970s by the US Bureau of Mines. But in the United States generally it is surface working for coal (strip mining) which poses the greatest potential problems of dereliction (see pp 166–8).

Related forms of mining

Much of what has been said about coal mining also applies to the forms of dereliction created by mining other types of stratified mineral. This is most obviously true of minerals worked from the Coal Measures, notably the two forms of ironstone – blackband and clayband – which were worked in several British coalfields before the 1940s. Although these ores formed the basis of the initial growth of the British iron industry between about 1750 and 1850, the low iron content of the thin seams which characterised most of the deposits was largely responsible for their abandonment, particularly when it became possible to use the equally lean but more cheaply worked Jurassic ores. In many instances coal and ironstone were worked from the same mine, so that the dereliction caused by working one of the minerals is indistinguishable from that caused by working the other. The mining of Carboniferous oil shales (Hoyle 1961) also produced types of dereliction similar to those found in the coalfields, notably in West Lothian, where the spoil heaps ('bings') composed of spent shale ('blaes') form a distinctive element in the landscape, which has been completely derelict since the demise of the industry in 1964.

Outside the coalfields the major bedded mineral to be worked in Great Britain is the lean Jurassic ore worked at different times since the 1850s from various horizons along the outcrop between Cleveland and north Oxfordshire. Much of the output comes from surface workings, particularly in the East Midlands (see pp 45–6), but mining was formerly important in Cleveland between 1856 and 1964, when the last pit closed. The Cleveland mines caused considerable damage through subsidence, for at their deepest they reached little more than 200m and the greatest thickness of ore (approximately 10m) was worked closer to the surface than this. In the East Midlands the Jurassic ores have only been mined sporadically, in the Scunthorpe area and in Northamptonshire. At Scunthorpe two mines, opened in 1939 and 1950 respectively, were responsible for one-fifth of the output from the Frodingham ironstone in 1965 (Pocock 1966). In Northamptonshire mining had greatly diminished in importance by the mid-1960s, with the closure of three mines, including that at Thistleton which had been opened as recently as 1956 in what subsequently proved to be a disappointingly lean sector of the orefield north of Corby. Abandonment has left some small pockets of subsidence-damaged land, which has, throughout the Jurassic outcrop, been the major form of dereliction produced by mining. Iron ore mines did not generally produce great mounds of waste, for the non-ferrous content of the mineral was not removed until the ore was smelted.

OTHER FORMS OF MINING AND DERELICTION

When minerals occur as irregular bodies, the techniques of mining may need to differ from those used in working stratified

Page 139 (*above*) The derelict Industrious Bee colliery, Ince in Makerfield. The site was being illicitly used as a rubbish dump; (*below*) The same view as preceding plate: an estate of local authority houses was built on the site after a reclamation scheme costing £1,570 per hectare

Page 140 Derelict land and reclamation near Wigan, Lancashire, 1971. This locality is depicted in Figs 35a and 35b

deposits, and the consequent forms of dereliction may also differ. In addition the physical and chemical properties of the mineral being mined, and the extent to which it is treated before dispatch to the refinery or smelter, may also create forms of dereliction which differ from those covered in the previous section of this chapter. A few examples will illustrate the variety of types of dereliction to be met with in mining areas.

Haematite iron ore mining in Cumberland and Furness

The haematite ores of Cumberland and Furness were a major source of raw material for British steelworks using the acid Bessemer process introduced in 1856, but even before that date the rich ore was much sought after by iron-masters. The haematite mainly occurs as irregular masses formed within the Carboniferous Limestone, either as 'flats', which are large lenticular bodies, or as 'sops', which were formed within sink-holes in the limestone. The deposits are thick and lie close to the surface in the exposed orefield, although they are almost everywhere blanketed with glacial drift. This set of physical characteristics, combined with the forms of mining most commonly employed, made damage by subsidence almost certain, and this, together with abandonment of plant, caused almost all the dereliction in the orefields.

The iron ore workings lie in three main areas, only one of which, the concealed orefield south of Egremont, had active mines in the 1970s. The principal mining area in West Cumberland during the nineteenth century was in the Ehen valley between Egremont and Frizington (Robinson and Wallwork 1970). The irregularity of the deposits meant that even in the heyday of mining there was a rapid turnover of pits, with much consequent dereliction. In the Ehen valley the mines were worked by pillar and stall extraction, taking out approximately 4m of ore at each level in the sop. When the limits of the mine were reached, the pillars were themselves worked, and after the collapse and compaction of the metalliferous debris, the ore was worked once again. The third mining district, Furness, comprised localities in Lancashire, between Askam and Lindal, and in Cumberland at Hodbarrow on the opposite shore of the Duddon estuary. The sops were worked in the same way, and the effect of this mining practice is well illustrated by the Park mine (Fig 25). Here a sop approximately 200m thick, 600m long and 200m wide at its broadest point, had sixteen shafts and four adits driven into it. The workings collapsed as mining progressed and ultimately, in common with many of the Furness mines, the site was permanently flooded. At Hodbarrow the ore body was a massive flat which was the most prolific source of haematite during the heyday of Cumbrian mining, producing about 500,000 tons of ore per annum, compared with averages of about 45,000 tons at most other mines (except Park, which had a peak output of over 300,000 tons). Hodbarrow was worked by the 'shrinkage' method, whereby horizontal sections 3m thick were extracted, the surface being allowed to collapse as each layer was worked. In order to work the undersea reserves by this method a sea wall was constructed, and when the original defences were undermined, another barrier was erected further out to sea. On abandonment of the mine in the late 1960s the pumps also ceased to work, and a lake formed over the site of the workings (Harris 1971).

Dereliction in the haematite orefields was not only caused by subsidence, although this generally produced the most spectacular damage. Spoil heaps were formed at each mine but, as the workings were relatively shallow and mined very pure ore bodies, heaps never became large and on abandonment most were rapidly colonised by vegetation. For this reason there is apparent disagreement between the 1965 1 : 25,000 Land Use Survey map of Dalton in Furness and the official returns of dereliction made by Dalton in Furness UD (in whose territory most of the old mines lie). The former classifies the majority of the mine sites as rough pasture or grassland, except where they are permanently flooded, whereas the latter records 300ha of derelict spoil heaps. This aptly illustrates the problem of determining when derelict land has appreciably blended with the surrounding landscape. It is perhaps indicative of the relatively inoffensive nature of the derelict land in Furness that none of it is scheduled for treatment in the Lancashire County Council's reclamation programme for 1971–6 (see pp 263–4). On the other hand the Cumberland County Council's programme for 1971–80 covers 200ha of derelict haematite workings, including 140ha at Hodbarrow.

Non-ferrous ore mining in Great Britain

Cornwall contains the greatest concentration of derelict land in England and Wales (see pp 92–7), much of it attributable to the mining of tin, copper and other non-ferrous ores. The mineral deposits occur as irregular veins and lodes intruded during the later stages of the igneous activity which formed the granites of the south-west peninsula. Until the mid-nineteenth century Cornwall was the world's principal source of tin; output reached its peak in the 1870s, declining dramatically after 1890 owing to foreign competition. Copper mining was also of great importance measured against world output, but after the peak years of production in the 1850s this too fell to severe overseas competition, notably from Chile and the United States. Dereliction began to appear a century or more ago in the Cornish mining districts, and gained momentum during the period from about 1895 to 1925. The Cornish mines resembled those in much of West Cumberland and Furness in scale of operation, with numerous shafts sunk at each mine to reach the irregular ore bodies. Often the shafts were sunk through dead ground, and galleries were driven from them to intercept the lode or vein. Most of the mines worked to depths of 400m or less, but exceptionally there were deeper shafts which were largely sunk to reach the tin lodes lying beneath the copper but separated from it by barren strate. Subsidence was rarely a problem, both because of the mode of working and the strength of the overlying strata, and it was only occasionally that 'runs of ground' occurred when the workings were close to the surface.

The major forms of dereliction were, therefore, waste heaps and abandoned industrial buildings.

The waste heaps comprised a variety of elements, including barren rock ('deads') removed during the sinking of shafts and the driving of underground roadways or trimmed from the ore at the surface. At copper mines ore dressing produced relatively little solid waste, but tin mines were obliged to adopt more elaborate processes which crushed the ore to a fine-grained material. The crushed ore was then concentrated and, if its chemical composition required it, the concentrate was roasted in calciners to drive off arsenical impurities. At each stage of treatment further waste was produced, much of it toxic. Most of the spoil heaps were manually tipped and in consequence were rarely more than 5m high. Only at the handful of mines which survived after 1920 were mechanical means of tipping adopted, but even so the resultant spoil heaps were mainly 10–15m high. Consequently the major long-term problem lay in their areal extent and toxicity. From time to time attempts were made to recover minerals from the mine dumps and the tailings at concentration plants, and there are still two works extracting tin from effluent in the Carnon and Red River valleys (Blunden 1970).

As there were 250–300 mines at work in Cornwall during the nineteenth century, only two of which survived in the 1940s, ruined mine buildings form the second major element of dereliction. They have evoked aesthetic responses not commonly associated with industrial decay, but not all the remains can claim to be scenically attractive in their own right. This is particularly true of those tin mines at which the massive foundations of the stamping batteries and the ruins of the calciners survive among the waste heaps and tailings. There are extensive groups of such buildings in the Red River valley at Tuckingmill and on the southern slopes of Carn Brea, both in the Camborne-Redruth area (Collins and James 1973). Revival of interest in Cornish tin mining potential during the 1960s (Goodridge 1966) does not seem likely to cause an increase in dereliction because the planning conditions attached to new mining ventures will be very stringent, and because the methods of processing ore at the surface are less wasteful than those employed in the past. Indeed the reworking of old mine dumps for recoverable metals may reduce the amount of derelict land more profitably than any other forms of reclamation.

Although Cornwall contains the major problem areas resulting from non-ferrous ore mining, the widespread nature of this industry in the nineteenth century means that many localities have pockets of similar forms of dereliction. Lead was mined in numerous upland areas, and in several of them substantial traces of the industry survive a century or more after all but a handful of the workings have been abandoned. In north and mid-Wales, for example, localities such as Halkyn Mountain, western Montgomeryshire and the Rheidol valley display numerous remains of lead mining in the form of abandoned shafts, ruined buildings and old spoil dumps (Thomas 1961, Edwards and Rees 1973). Similar derelict landscapes appear in the Pennine lead-working districts, from the northern dales through Craven (Raistrick 1970) to the Peak. In the last-named area considerable surface disfiguration survives from the nineteenth century, comprising ruined mine buildings and smelting plants, together with spoil heaps and deep trenches following the outcrop of the lead rakes (Carr 1965).

One problem posed by almost all the former non-ferrous mining areas is that they lie in areas of great amenity value, often designated National Parks. While this offers great incentives to reclaim the derelict sites, it also means that a revival of interest in the mining potential of such localities is likely to produce a major clash over land-use priorities. New forms of mining present a similar dilemma when they are in conflict with visual amenity, as is well illustrated by the siting of potash mines in the North York Moors National Park. Although the three proposals to mine potash have been subject to the most stringent planning controls, the fact remains that problems of waste disposal and the risk of subsidence both exist. While most of the arguments against this mining development centre on its intrusiveness during the period of active exploitation (Statham 1971), the possibility of creating future dereliction cannot be entirely ignored.

Rock-salt mining in Cheshire

By virtue of its chemical composition rock-salt poses a particular difficulty when mined, in that the workings have to be kept completely dry in order to prevent accidental solution of the mineral. (The same point applies to the mining of potash.) Inability to achieve this was the principal cause of collapse at most of the rock-salt mines worked in the Northwich area of the Cheshire saltfield during the period between 1670 and 1928 (Wallwork 1956, 1959, 1973). In this locality the major rock-salt deposits comprise two beds, each approximately 30m thick, separated by a band of Keuper Marl; the uppermost bed lies at a depth of about 70m below the surface. This combination of thick deposits and shallow workings produced the optimum conditions for mining subsidence, and as almost all the mines were situated on the floodplain of the River Weaver or Witton Brook, the risk of water entering the workings was increased.

The rock-salt was mined by the pillar and stall method and massive columns, 4–6m^2 in plan, were left to support the roof of the galleries from which a thickness of 4–13m of rock-salt was extracted. Occasionally the pillars were inadequate to bear the weight of the overlying strata, causing subsidence above the mine, but the most common cause of collapse was erosion of the pillars by water which had accumulated on the sole of the mine. Between 1750 and 1881, the major period of mining subsidence, approximately fifty mines were flooded; a further eight had been abandoned by 1915 and the last, the Adelaide mine, was flooded in 1928. The dereliction caused by the collapse of rock-salt mines took the form of funnel-shaped 'rock-pit holes'. Subsidence was confined to the vicinity of the collapsed shafts and as the floodwaters became supersaturated brine, no further solution occurred. For many years the rock-pit holes remained as small lakes, rarely more than 10m across, but during the 1880s much more severe subsidence caused by brine pumping obliterated most of them (see pp 151), though one or two of the lakes produced by later mine collapses still survive (see plate, p 157). Rock-salt mines produced very little solid waste, so that spoil heaps were never a characteristic of the mining landscape. The mine buildings were generally flimsy structures, and the collapse of the shafts almost always meant that they disappeared with the abandonment of the mine, unlike the solid engine-houses of the non-ferrous ore mining districts. Consequently hardly any of the dereliction in the Cheshire saltfield is directly attributable to rock-salt mining, although much of the worst damage occurred in the mining district at Northwich when the abandoned workings were tapped as reservoirs of 'bastard' brine (see p 275).

Other forms of mining

The range of examples cited above spans the commonest forms and causes of dereliction associated with mining, particularly in Europe. In the New World the early stages of mineral exploitation produced similar forms of disruption to those found in nineteenth-century Europe. Thus the reports of the United States Geological Survey during the second half of the nineteenth century contain numerous descriptions of the small-scale mines which worked the minerals of the Rocky Mountains and the intermontane basins and plateaux. Even though the individual mines might be small, there were many instances in which the concentration of mining in a relatively small but geologically rich area produced major disruption and disfigurement at the surface. The Klondike goldfields provided an example of this in the 1890s, for they were worked by numerous small shafts sunk to depths of 30–60m which could be picked out long after their abandonment by the lines of spoil dumps along the floodplains and valley flanks of the various creeks (Meiklejohn 1900). Many of the world's goldfields bore a similar appearance during the nineteenth century, with added damage from equally primitive forms of surface working and mineral treatment. Indeed the desolation of the abandoned mines and their associated 'ghost towns' has become an important element of the folk-history of western America and of similar areas in Australia and New Zealand.

However, during the twentieth century the most important feature of most of the non-ferrous ore mining areas outside Europe has been the tendency to concentrate working at a handful of very large mines, with the probability that the amount of potential dereliction will be proportionately greater. Thus in the lead-mining district of south-east Missouri the workings comprised a large number of shallow shafts before the perfection of deep mining techniques in the 1870s. Thereafter mining was concentrated at a smaller number of shafts each marked at the surface by massive dumps of waste tailings. However, one of the greatest contrasts between old and new methods of mining, with important consequences for the creation of potentially derelict land, is provided by gold mining. The ephemeral shallow-shaft mining which was typical of the early gold workings of North America and Australasia was not repeated in the modern development of the Witwatersrand goldfield. Here the gold reef lies at depths of as much as 3000m and has had to be worked from deep shafts. The initial processes of ore treatment take place at the mine and some of the sand from the reduction plants is returned underground for use as back filling. Much of the waste remains at the surface in slime dams and in huge flat-topped dumps of white sand, which are being added to at a rate of about 165,000 tons per day. Although the dumps have been worked for uranium since 1949 (Niddrie 1955), they continue to form a major element of the industrial landscape along the mining zone, which stretches for 95 km through the Witwatersrand conurbation (Cole 1957).

FLUID MINERAL EXTRACTION AND DERELICTION

The abstraction of minerals in fluid form may produce similarly damaging side-effects to those caused by conventional mining, notably subsidence. It is far less common for mineral working of this kind to produce solid wastes, although pollution of the atmosphere and of watercourses may result from the disposal of waste residues, and the initial stages of refining may yield solid waste. Fluid mineral extraction takes three principal forms:

1 tapping natural seepages of minerals;
2 pumping natural occurrences of mineral fluids to the surface;
3 solution mining (artificially creating fluids from soluble solid mineral deposits).

The principal minerals abstracted by one or more of these methods are fossil fuels – oil and natural gas – and soluble solids, including sulphur and salts such as potash and sodium chloride. The fossil fuels are always worked in liquid form, but the soluble minerals may also be worked as solids by conventional methods of mining. In addition to the abstraction of fluid minerals, excessive pumping of ground water may also lead to subsidence, as has happened in Mexico City, central California, the Houston-Galveston area of Texas and Greater Tokyo, to name the four most extreme examples (Legget 1973).

Oil and natural gas subsidence

The extraction of crude oil and natural gas has caused identifiable subsidence in fields at or near sea level, where it is most readily observed. It has almost certainly occurred in other areas where its effects are either immeasurable or immaterial. Undersea oil and gas fields, or those located in barren areas, are not likely to display the same damaging effects of subsidence as those in, say, the Po delta or on the shores of Lake Maracaibo.

The subsidence is caused by reduction of fluid pressure in the oil and gas zones, which leads to settling and compaction of strata. The most noteworthy instance of this type of subsidence has been recorded in the Wilmington oilfield, Long Beach, from about 1937 onwards. This subsidence is well documented because it has affected an intensively urbanised area lying 2–3m above sea level. Between 1940 and 1962 the centre of the subsiding area sank approximately 8m, and there was an accompanying horizontal displacement of up to 3m during the same period. Expensive remedial measures, costing over $100 million during this period, were undertaken in order to protect oil production and to prevent extensive dereliction (Poland and Davis 1971). From 1958 onwards, however, the remedial measures were accompanied by attempts to arrest and reverse subsidence by injecting water under pressure into the oil zone. Repressurisation has produced a surface rebound of up to 15 per cent of the original subsidence, but the expense of these measures is only likely to be borne in wealthy urban-industrial areas such as the Los Angeles city region.

Brine-pumping subsidence

In solution mining technology the term brine is applied to the liquid form of all soluble salts, but in this instance it relates solely to sodium chloride (common salt) and its extraction in the Cheshire saltfield, which currently accounts for over 80 per cent of the United Kingdom's salt production. Brine occurs naturally in those saltfields where circulation of ground water has dissolved the upper layers of rock-salt, forming a dense layer of brine termed the wet rock head. Although it is evident that natural subsidence has occurred on a large scale since the salt beds were laid down in the Triassic, the formation of derelict land is a product of less than two centuries of human exploitation. Initially the salt-refining industry relied on natural seepages of brine at springs, and little or no observable subsidence took place. During the eighteenth century commercial pressures and technical improvements caused the salt trade to expand, the increased supply of brine being secured by pumping from natural brine reservoirs at depths of 70–130m (Wallwork 1959). This caused accelerated subsidence from about 1790 onwards. Its precise form and location varied according to the position of the brine runs, the strength of the overlying strata, the topography of the subsiding area, and the disposition of the pumping shafts (Wallwork 1956). Suffice it to say that from 1790 to the present day the pumping of brine from the wet rock head has caused considerable dereliction in the Cheshire saltfield (Wallwork 1960a, 1960b, 1973), and it continues to pose major planning problems in the locality (see pp 272–5).

The damage caused by pumping natural brine generally occurs slowly and is unpredictable in any locality underlain by the wet rock head. During the period 1873–1935 much spectacular and rapid subsidence took place in the area underlain by abandoned rock-salt mines at Northwich. The floodwater in these workings was supersaturated brine, which, when pumped out, permitted the entry of fresh water and the rapid destruction of the remaining pillars of rock-salt within the abandoned mines. There followed a series of spectacular subsidences which were recorded in great detail by contemporary observers (Hull 1883, Ward 1898, 1900) and can also be measured from Ordnance Survey maps and plans of the period (Wallwork 1956, 1960b). Extensive and sudden mine collapses, the progressive expansion of the 'flashes' (subsidence lakes), and the formation of fringing swamps and badly broken ground transformed what had been a prosperous salt-mining and refining district into a wasteland within little more than 30 years when 'bastard' brine pumping was at its peak. That this tract of land became valuable as a dumping ground for chemical waste from the adjacent ammonia-soda plants (Wallwork 1960a, 1967, 1973) scarcely alters the fact that the original damage represented a highly concentrated and singularly rapid growth of dereliction caused by the extraction of a fluid mineral (see Fig 37).

One reaction to this violent disruption and disfigurement was to investigate means of securing brine supplies without fear of interruption from subsidence. During the 1920s the introduction of a method of solution mining (termed 'controlled pumping') both avoided the uncertainties of supply encountered in tapping natural brine runs and appeared to minimise the risks of subsidence. The technique involves drilling boreholes to rock-salt beds which are below the zone of ground water circulation.

Water is forced down the borehole and the resultant brine flows to the well head under pressure. The cavities which are formed in the rock-salt are separated by massive supporting pillars and, as they are situated well below the wet rock head, there is little fear of accidentally tapping natural brine runs and thereby causing further subsidence.

Controlled pumping has at least two other virtues. Firstly, the pumping shafts occupy only 3 per cent of the surface area of the Holford brinefield, so that farming can proceed normally above the workings. Secondly, it has proved possible to use the worked-out cavities as receptacles for chemical waste which had formerly been dumped either on derelict land or, increasingly during the 1940s, on sound agricultural land. Since the 1950s the Holford brinefield has been able to accommodate both mineral abstraction and waste disposal without experiencing the once normal disruption and disfigurement. Unfortunately controlled pumping cannot be employed in areas which are already experiencing subsidence from conventional methods of brine pumping and, for as long as natural brine runs are tapped, subsidence and its concomitant forms of dereliction are bound to occur (Collins 1971).

Mining and dereliction

That mining causes a substantial part of the dereliction recorded throughout the world is undeniable, but it is also clear that there is no simple solution to the problems presented by derelict land in mining areas. The great variety assumed by dereliction is in itself an indication of the complexity of the problem. Geology and topography, the techniques of mining and mineral preparation, the law relating to mineral rights and the physical restoration of damage caused by it, and the economics of mineral production are the major variables which in combination explain the origins and development of dereliction. The significance of the various types of derelict land to those faced with the tasks of reclamation and restoration is examined in Chapters 7 and 8. Two general observations about the nature of dereliction should, however, be made at this point. Firstly, the technical means to minimise or even to avoid the creation of further dereliction often exist but are not used, or at best are underemployed, because the cost of universally introducing them might reduce the competitiveness of a particular mineral working in the open market. The absence of rigorous cost-benefit analysis means that the social and economic costs of derelict land never enter the equation when mining companies argue that remedial techniques are too expensive to introduce. Few if any of the methods of mining and waste disposal which have helped to minimise the amount of dereliction were conceived with that end solely in mind, for it has been much more common for technical and commerical considerations to dominate decision-taking, with the reduction of dereliction as an incidental bonus. Secondly, dereliction caused by mining is not normally a product of the early primitive phases of the mining industry's development. If the problems of dereliction were confined to areas worked during the nineteenth century alone, they would generally be more easily solved. It is no accident that the forms of derelict land which many people find least offensive are those dating from the earliest stages of mining. Cornish engine-houses 'perched like the nests of eagles' on the Atlantic coast, or Alaskan gold diggings immortalised in ballad and legend, do not fall into the same category as the mountainous spoil heaps which straddle the coalfields of western Europe. Dereliction is, therefore, a continuing feature of the evolving landscape in mining areas, and there seems little likelihood that it will cease to be so in the immediate future.

References to this chapter begin on p 310.

CHAPTER FIVE

Surface Mineral Working and Dereliction

SURFACE mineral working includes all forms of mineral extraction carried out at excavations open to the sky. In common English usage there are three words which describe such mineral workings, but with a lack of precision which in two instances leads to confusion with mining as defined in the previous chapter. 'Quarry' normally means an open excavation for building-stone, slate, or other 'hard' rock with no metalliferous or combustible content. 'Pit' is used to describe open workings in 'soft' or unconsolidated deposits, such as clays, sands and gravels, and originally related to their shallowness. 'Mine' is used to describe excavations working fossil fuels and metalliferous ores, frequently with a qualifying adjective, eg opencast or open-cut mine, strip mine, patch mine. Both pit and mine are terms also applied to underground mineral workings, but their use in this chapter is confined to surface workings according to the appropriate common usage. Where it is necessary to refer to underground mineral workings, the term 'deep mine' will be used.

Forms of dereliction

Dereliction caused by surface mineral working mainly stems from three characteristics:

1 the amount of overburden which has to be stripped before the mineral is reached;
2 the ratio of usable mineral to waste removed at the workings;
3 the areal extent and depth of the excavation.

The form taken by the workings and by the consequent dereliction is a compound of physical and technical influences, the latter mainly reflecting developments in the design of excavating and earth-moving machinery. The problems of waste disposal resemble those encountered at deep mines, but with the added complication that waste may also accumulate on the floor of the workings as well as at the cleansing and preparation plants. For example, the principal form of dereliction left by opencast working of the Jurassic ironstones in the East Midlands before 1951 was 'hill and dale' deposited behind the advancing quarry face as mechanical excavators dumped the overburden. In contrast, the waste at china clay pits in Cornwall and Devon lies alongside the workings in the form of white pyramidal dumps of micaceous sand. Dereliction caused by abandonment of the workings themselves may leave dry pits and quarries or permanently flooded ones, according to the geology, topography and level of the water table at individual sites. In addition dereliction may be caused by the abandonment of surface plant and of ancillaries such as settling ponds and mineral railways, as with other forms of mineral working and processing.

In analysing the forms of derelict land associated with surface mineral working S. H. Beaver (1944, 1961) designed a system of classification which, with amendments devised to cover a wider range of examples, forms the structure upon which this chapter is based. The categories recognised, and the examples used to illustrate them, are indicated below. Some minerals have been worked at more than one type of excavation over the years and recent developments in earth-shifting technology have tended to blur the former distinctions between the various categories.

1 Excavations working thin mineral deposits (6m or less) with little or no overburden and not reaching the water table. (Most dry sand and gravel pits and all peat diggings fall into this category. In earlier periods of development opencast workings for ironstone, coal and clay were also of this type.)
2 Excavations working thin to moderate mineral deposits (20m or less) with overburden reaching thicknesses of 30–35m (opencast mining of coal and ironstone).
3 Excavations working thick mineral deposits with varying amounts of overburden. The following sub-types are identifiable:
   (a) working 'soft' material which can normally be excavated without blasting (brick and china clays);
   (b) working 'hard' material which normally has to be blasted before excavation (slate, limestone, chalk);
   (c) working fossil fuels (brown coal or lignite);
   (d) working metalliferous ores, which may be either 'soft' or 'hard' and, unlike most examples in (a) and (b) above, may also be worked by deep mining techniques in other localities (metalliferous ores).
4 Surface working of deposits taken from below the water table (but excluding off-shore dredging):
   (a) 'pure' workings (sand and gravel);
   (b) 'residual' workings (gold).

TYPE 1: SHALLOW WORKINGS

Shallow surface workings normally create few long-term problems of dereliction, for they can be reclaimed cheaply or may rapidly blend with the landscape after abandonment, provided that they do not become waterlogged. They were most numerous in the early exploitation of minerals by surface working and in many instances little or no trace of them remains. There is also scant record of them in official statistics: British reports ignored workings less than 6m deep, which were not covered by legislation relating to the operation of mines and quarries. During the nineteenth century many towns supported their own brickworks, using a variety of clays; often

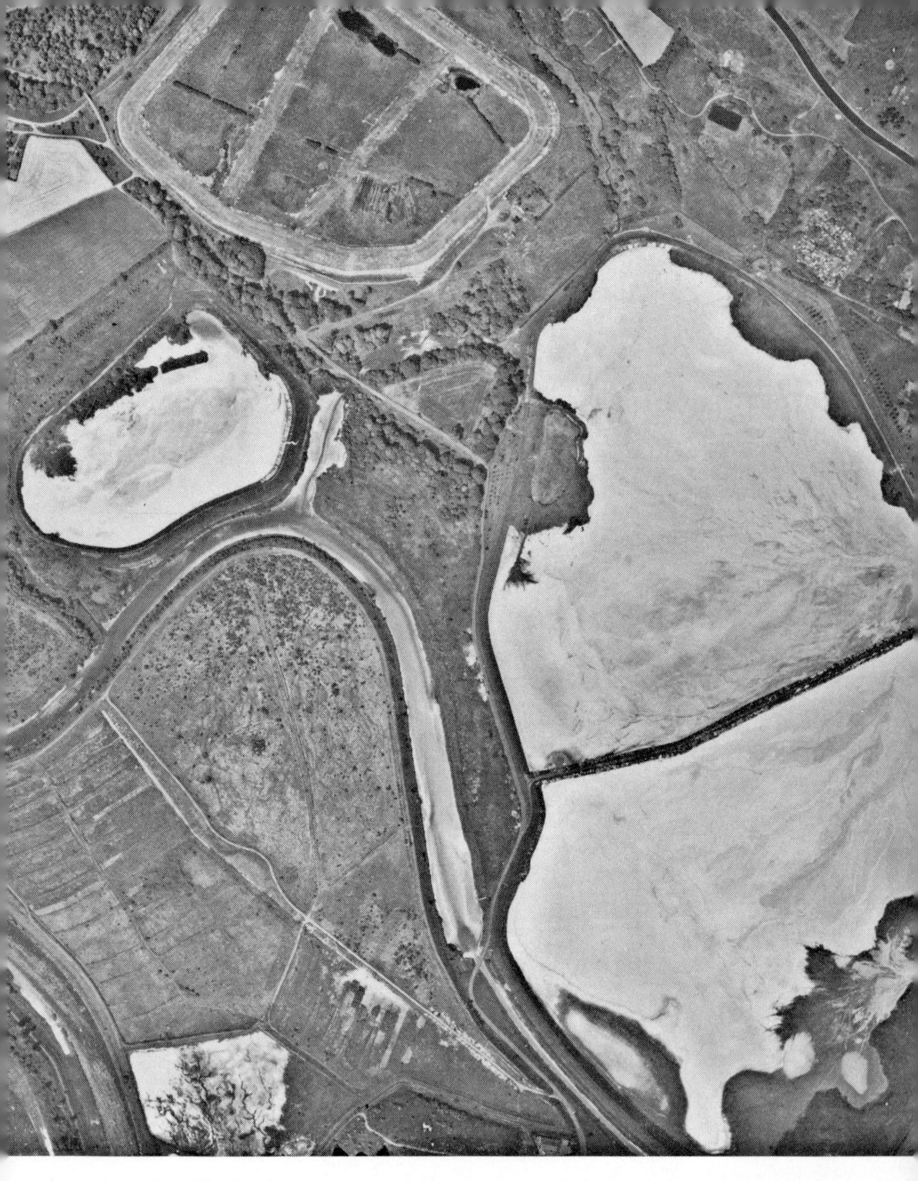

Page 157 Derelict land near Northwich, Cheshire, 1967. A nineteenth-century salt refining and rock-salt mining district, severely damaged by subsidence and subsequently used for dumping chemical waste. The lime bed on agricultural land (*top left*) was constructed in 1953, but remained unused as waste could be disposed of below ground (see also Figs 2 and 37)

Page 158 Industrial land use, Lostock Gralam, Cheshire, 1967. Lime beds in various stages of use lie close to the ammonia-soda works. The remains of solid waste from a Le Blanc acid and bleach works are also to be seen, together with the sites of abandoned brickfields (see Fig 28c)

all that remains is, at most, a hummocky meadow or a street or district name indicating the former existence of brickfields. Similarly, sand and gravel normally came from ephemeral shallow pits serving a local market. For example, the terrace gravels and brickearths of the Lea Valley in Middlesex were extensively worked between 1870 and 1920, but most of the sites of these dry workings have been built over or converted into public open spaces. Where sand and gravel is now worked from shallow dry pits, the likelihood of long-term dereliction is equally slight. The sand and gravel industry in Great Britain has gained a good reputation for sponsoring schemes of reclamation; even though pits up to 40ha in extent are commonly worked, disruption only occurs during the period of active working and the workings are normally restored after use.

In the British Isles the only other extensive shallow diggings are for peat, which is mainly worked for horticultural use. The most important English workings are in the Somerset Levels, where peat has been worked commercially since 1870, but a greatly increased rate of extraction since 1954 has given rise to the threat of dereliction (Dale 1967). The peat lies beneath a thin cover of top soil and at its maximum is about 6m thick. Most of the workings do not yet exceed 3m in depth and are above the water table, but the need to achieve greater output is likely to be met by removing all the peat to its junction with the underlying blue clay. Much of the early peat digging was confined to the top metre or so and the abandoned workings are now covered with birch and wet-heathland vegetation. Diggings begun since 1958 have been subject to more stringent planning controls which, *inter alia*, require the restoration of the workings to agriculture and the protection of the delicately balanced drainage systems. However, the life of individual workings may be as long as 25 years and large-scale reclamation has yet to take place. Without adequate control the extraction of peat could ultimately create 500ha of flooded land which might be difficult to restore or adapt for amenity use.

Other major areas of commercial peat working occur in the

Republic of Ireland, where approximately 160,000ha of peat bog may ultimately be worked to depths of 3–4m in order to provide fuel, industrial raw materials and horticultural products. By the early 1960s about one-third of this area was already being exploited and experimental work was being undertaken to ensure that excessive dereliction did not accompany the abandonment of the workings, most of which have a life expectancy of 20–25 years (Dwyer 1962). An unknown proportion of the 600,000ha of previously cut-over bog in Ireland already lies derelict. In Northern Ireland, where small-scale piecemeal exploitation has been the norm (Johnson 1959), some of the major bogs are at least three-quarters derelict, according to an official survey carried out in 1956. However, it is the major commercial workings which pose the biggest potential problems of dereliction. Bord na Mona (the Turf Development Board) has already initiated pilot commercial projects based on the reclamation of bog land for agriculture, horticulture, and forestry at three locations in east-central Ireland (Cluain Sosta, Loilgheach Mor, and Doire Dhraigneach). It remains to be seen whether all the commercially exploited bogs will be restored to other uses: if not, they could be included among the largest tracts of derelict shallow workings in Europe (Coudé. 1973).

TYPE 2: DEEP WORKINGS WITH A HIGH RATIO OF OVERBURDEN TO MINERAL

Surface mineral workings of this type only became possible with the advent of high capacity earth-moving machinery which could strip overburden from workable deposits at low cost. Although such equipment was first introduced to Great Britain in the 1880s, it was not widely employed before the 1930s. An important characteristic of surface mineral working since World War II has been the rapidity of technical improvements in excavating and earth-moving machinery. Among other things this has increased the profitable ratio of overburden to mineral, and by implication the economic depth of working. For example, when opencast coal mining was reintroduced to Great Britain in 1942, the acceptable ratio of overburden to workable coal was 4 : 1; 30 years later it was 40 : 1.

Surface working of ironstone and coal in Great Britain

Opencast mining of ironstone and coal was a feature of the early stages of working both minerals in many parts of Great Britain, but as most of the excavations were shallow, little long-term dereliction stemmed from their abandonment. The principal exceptions to this are provided by many localities along the north crop of the South Wales coalfield between Pontypool and the Gwendraeth valley, in which 'scouring' and 'patching' were employed to work both coal and ironstone. Scouring involved removing the overburden by hydraulic sluicing; patching was a primitive form of shallow opencast mining. Together these methods laid waste large tracts of land between the 1750s and the 1920s, evidence of which is still to be seen in the roughly vegetated gullied landscapes of the north crop (see plate, p 68). Such workings are the linear antecedents of modern ironstone and coal opencast mines: indeed in several instances the abandoned patchworks have been reclaimed after modern opencast reworking of their sites.

Almost all the opencast working of ironstone in Great Britain since the 1850s has been confined to the Jurassic outcrop of the East Midlands. The ironstone deposits south of the Humber are being, or have been, exploited in four principal localities. In the Scunthorpe area the Frodingham Ironstone (Lower Lias) yields an exceptionally lean ore (20 per cent iron content), which accounted for roughly 37 per cent of Jurassic output in 1969. Further south the Northampton Sands Ironstone (Inferior Oolite), with an iron content of 31 per cent, accounted for 61 per cent of Jurassic output in 1969. The two remaining localities, north Oxfordshire and Leicestershire, lie on the Marlstone deposits of the Middle Lias, which in 1969 produced only 2 per cent of the Jurassic ore, compared with about 20 per cent a decade previously. Working ceased in north Oxfordshire in 1967 and the Leicestershire field was also diminishing in importance by this date (Wheeler 1967).

During the earliest phases of opencast mining at Scunthorpe, from 1859 to 1904, the quarries were extensive but shallow (Pocock 1964). The amount of overburden was slight, being nowhere more than 1m thick, and only the top 3m or so of the ore was taken. Quarrying was entirely by hand, although machinery was introduced in 1885 to strip the overburden. Between 1904 and 1919 the pattern of quarrying changed appreciably as working moved down dip and it became necessary to remove more overburden. In 1904 more efficient stripping machinery was introduced, and in 1912 mechanised ore excavation began to replace manual quarrying. By 1918 up to 18m of overburden was being removed, and the ironstone itself was being worked to thicknesses of 10m. In part this was achieved by quarrying to greater depth in abandoned shallow workings, or by quarrying the sites of disused ironworks and their slag heaps. Since the 1930s increasing amounts of overburden have had to be removed, and by the 1960s over 40m was being stripped in the Roxby quarries at the foot of the Oolitic escarpment. In spite of this lengthy history of ironstone working, the Scunthorpe area, in common with the Jurassic orefield generally, contains very little derelict land, though partly for different reasons. Roughly two-thirds of the workings lie within the municipal limits of Scunthorpe, where much of the quarried land has been used for industrial building. In addition, many of the early shallow quarries which were not reworked lay in areas of poor quality land, so that reclamation posed fewer problems than in the richer agricultural areas further south (Fig 26).

On the Northampton Sands orefield the potential for dereliction was greater. Early workings were shallow, with relatively little overburden, and even after the introduction of mechanised stripping at Corby in 1895, the ore was dug and the overburden replaced and levelled by hand (Pocock 1961). The workings were, however, more numerous than at Scunthorpe and much more widespread over an area of high quality farming. As working

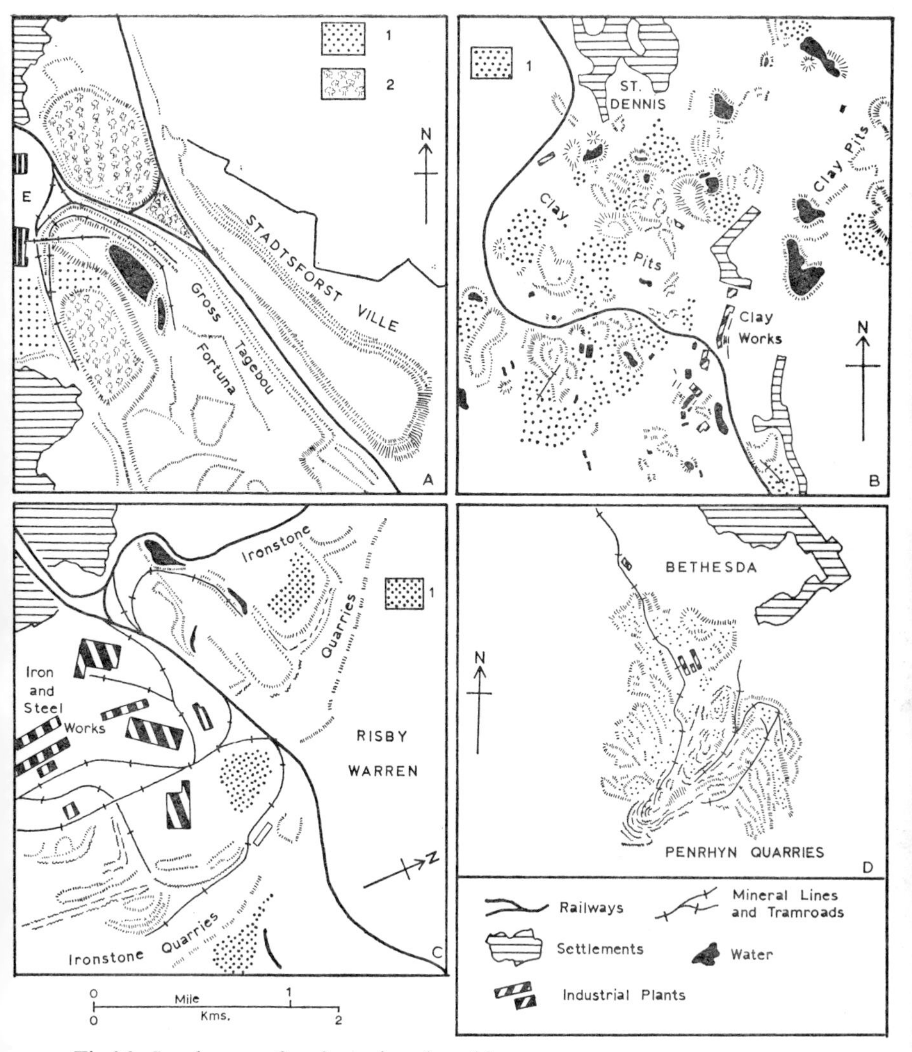

Fig 26 Landscapes of surface mineral-working areas
A The Cologne brown-coalfield near Frechen: 1, reclaimed land; 2, re-afforested land; E, electricity works
B China-clay workings north of St Austell, Cornwall: 1, waste heaps
C Ironstone quarries, Scunthorpe, Lincolnshire: 1, overburden and waste heaps
D Slate quarries, Bethesda, Caernarvonshire
All the maps are to the same scale. Industrial plants generalised

moved down dip during the 1920s and 1930s, it became necessary to remove increasing quantities of overburden. With the introduction of modern machinery in the 1930s, the practice was established of dumping the overburden behind the advancing quarry face, thereby creating 'hill and dale'. By 1938 approximately 1,200ha of hill and dale had been created in the East Midlands, and as the workings continued to proceed down dip, it seemed inevitable that this land-use problem would grow in scale. That it did not was due to the findings of the Kennet Committee in 1938–9 and the provisions of the Mineral Workings Act 1951. The Kennet Committee considered that the reclamation of hill and dale should fall as a financial obligation upon the mineral operators through a levy on tonnage raised (see pp 242–3). It was thought that afforestation of hill and dale, as practised in parts of Northamptonshire before 1939, offered the most feasible means of reclamation, and that this would in turn provide pit props when it became necessary to deep-mine most of the ore. In fact this necessity never arose because the massive dragline excavators introduced to the Corby area in the 1950s can strip 30–35m of overburden from the ironstone. Even more important is the fact that it has proved possible to level much of the existing hill and dale (see plates, p 104) and to prevent its formation in newly worked areas (see plate, p 50). Neither Northamptonshire, with its long history of ironstone working, nor the worked-out Marlstone fields of north Oxfordshire and Leicestershire contain much derelict land.

Opencast coal mining was reintroduced to Great Britain in 1942 as a short-term emergency measure. In spite of various statements made since 1945 that this 'temporary expedient' was about to end, opencast coal production continues, although at a relatively low rate to protect the deep-mining sector of the industry. Its continuation reflects the profitability of opencast mining, which permits a much higher proportion (approximately 90 per cent) of the coal to be worked than is possible with deep mining (Bennett 1969). The earliest opencast working was normally confined to outcrops at a depth of less than 15m, with an overburden to coal ratio of about 4 : 1. During the 1950s advances in mining technique made it possible to work to depths of 100m, with a ratio of 20 : 1, and in the late 1960s opencast coal workings were being excavated to depths of 150–230m, with a ratio of 40 tons of overburden to 1 ton of coal. In many instances individual sites have been worked at different depths over a period of time, as in East Warwickshire. Elsewhere it has been possible to work opencast coal down dip from the earliest workings at depths that would not have been thought possible in the 1940s.

Although 290 million tons of coal had been won by opencast working in the years 1942–72 (Table 8), this form of coal

TABLE 8 Opencast coal mining in Great Britain (million tons)

1942	1·3	1953	11·9	1964	6·9
1943	4·5	1954	10·3	1965	7·4
1944	8·8	1955	11·6	1966	7·1
1945	8·2	1956	12·3	1967	7·2
1946	8·9	1957	13·8	1968	7·0
1947	10·4	1958	14·6	1969	6·4
1948	12·0	1959	11·0	1970	7·9
1949	12·7	1960	7·7	1970*	6·6
1950	12·4	1961	8·7	1971*	8·3
1951	11·2	1962	8·2	1972*	10·4
1952	12·4	1963	6·2		

* NCB Financial Year; other data are for calendar years
Data from *Annual Abstract of Statistics* and NCB *Annual Reports*

mining has produced no long-term dereliction. The disruption caused during the period of active coaling is often spectacular, but it has always been official policy to restore the sites after extraction. This was simply achieved by virtue of the emergency regulations under which opencast mining was introduced, and was aided by the fact that mining was carried out by civil engineering contractors who developed techniques of restoration using the machinery normally at their disposal. Even more important is the fact that the Opencast Executive of the NCB has in recent years coordinated its programme of mining with other major schemes devised to reclaim derelict land (see pp 236–7). Consequently opencast coal mining in Great Britain provides an excellent example of mineral working which has produced no long-term dereliction, a position shared by ironstone quarrying in the East Midlands since 1951.

Strip mining of coal in the United States

Strip mining accounted for 36 per cent of anthracite, lignite and bituminous coal output in the USA in 1965 (compared with 4 per cent of British coal production attributed to opencast working). The adoption of widespread strip mining during the 1940s reflected technical developments in the design of excavating machinery, but the same constraints were not placed on this form of working in the USA as were applied in Great Britain. Three types of strip mining have been employed:

1 area mining;
2 contour mining;
3 hybrid area-contour mining.

Area mining is normally carried out on level or gently sloping terrain. The process entails cutting parallel trenches to expose the coal seams, each trench being the receptacle for overburden removed from the next in the succession. This produces a 'washboard' landscape with ridges up to 15m high, which is commonplace in the strip-mined coalfields of Illinois and Kentucky. Contour mining is largely confined to mountainous areas and is particularly well developed in the Appalachian coalfields. In this method of working a bench is created along the coal outcrop, bounded on the upslope side by a sheer face of solid rock which may be as high as 30m, and on the downslope side by a precipitous spoil-covered slope which may be excessively unstable. The hybrid form is found in the Pennsylvanian anthracite fields, where the complexity of geology and topography may produce either extensive quarries or canyon-like trenches. In addition auger mining (using drills which scoop out the coal) is carried out to reach those deposits which cannot be reached by strip mining.

In 1965 it was estimated that 526,681ha of land had been disturbed by strip mining for coal in the USA, or 41 per cent of all land damaged by various types of surface working. Pennsylvania contained the major share (23 per cent), followed by Ohio (16 per cent), West Virginia (15 per cent), Illinois (10 per cent), and Kentucky (9 per cent). In all five states coal strip mining was the primary source of damage and no other form of dereliction exceeded 50,000ha in any state of the union. The scale of disruption has been catalogued by the United States Department of the Interior (USDI 1967), which estimated that about two-thirds of all land disturbed by surface mineral working remained derelict. Unfortunately no breakdown of this statistic is available by type of mineral working, but it is provided for each state. In the five states listed above 61 per cent of the disturbed land had not been reclaimed and, as coal strip mining accounted for 88 per cent of all their disturbed land, it is reasonable to suppose that a little under two-thirds of the strip-mined land remained derelict. Data have not been collected in the US since 1965, but it was estimated that an additional 20,000ha of strip-mined land would be created each year at 1965 rates of production.

Although some impressive schemes of reclamation have been carried out, the fact remains that this form of working has been far more disruptive in the USA than in Great Britain. Thus it was estimated in 1965 that there were 40,000km of contour mine benching, about three-quarters of it bounded by dumps of overburden; along the remaining length spoil had been tipped downslope. This was not only unsightly but also potentially dangerous, for the instability of the waste heaps and the lack of vegetative cover led to excessive slumping, blockage of water-courses and in extreme cases mass movement of unconsolidated waste. It must also be noted that much of the worst damage is concentrated in Appalachian mining valleys with long-standing social problems that were exacerbated by the loss of employment in deep mines caused by the advent of strip mining (Udall 1966).

TYPE 3: DEEP WORKINGS WITH LITTLE OVERBURDEN AND THICK MINERAL DEPOSITS

It is possible to identify four major forms of mineral working within this category, each of which is capable of producing extreme forms of dereliction. In some respects it has become increasingly difficult to separate Type 3 workings from the deeper Type 2 workings discussed above, the major distinction being that Type 3 workings have a very low ratio of overburden to mineral thickness and can be worked economically to much greater depths. Six examples will be considered in this account, in order to illustrate some of the principal problems of extant and potential dereliction.

(*a*) *Deep workings in 'soft' material*

(*i*) *Brick clay in Bedfordshire*. Before the 1920s, brick clay was commonly worked at numerous shallow pits located on a variety of geological deposits. In 1939 there were over 1,250 brickworks in Great Britain, but their number had fallen to 525 in 1958 and to 324 a decade later, partly owing to fluctuations in the demand for bricks and partly to the increasingly successful competition of a handful of large and efficient plants in the Jurassic Oxford Clay vale. In 1948 the production of 'Fletton' bricks from this area accounted for 31 per cent of British brick output, and by 1966 the proportion had risen to 42 per cent. The second most important source of bricks is the Carboniferous Coal Measures clays, with a much greater spread of locations in various coalfields.

The value of the Oxford Clay as a source of brick clay was first recognised in 1881 when works were established at Fletton, Peterborough. Oxford Clay can be fired immediately without weathering and, as it contains 10 per cent of carbonaceous material, needs only one-third the amount of fuel required to fire other clays. Low fuel costs coupled with a favourable location and adoption of mechanised methods of manufacture from 1896 onwards enabled brickmakers in the clay vale to capture a much wider market than was normally served by brickworks. This in turn attracted further capital investment and the establishment of larger concentrations of brickworks. The first brickworks to use the Oxford Clay in Bedfordshire was established in 1897, but not until the inter-war years did the major plants with their massive clay pits (locally termed 'knott holes') begin to dominate the scene. The principal firm, founded in the 1930s after a series of mergers, is the London Brick Company, which by 1971 controlled all brick manufacture in the vale with the exception of works at Whittlesea, Cambridgeshire, operated by a subsidiary of the NCB (Cowley 1972).

In Bedfordshire the workable clay 'knotts' lie beneath a worthless brown clay termed 'callow', which is between 3 and 12m thick; in addition a non-carbonaceous blue clay may also overlie the knotts, but attempts to employ this in brick making have not been successful. The callow and blue clay rejected during the working of the pits form the ridges of waste on the floor of the abandoned excavations (see plate, p 67). The workable clay averages a thickness of 14–16m, so that given maximum overburden the pits may reach a depth of 30m. Dereliction in the brickfield comprises three main elements: dry pits with ridges of waste, flooded pits and abandoned kilns, together with ancillary plant. In 1972 the London Brick Company controlled 1,678ha of clay covered by planning permissions to excavate, in addition to 627ha which had either been worked and abandoned or were currently in use (Fig 27). Much of the worked land was subject to after-use conditions laid down in 1947–9 and subsequently considered to be wholly inadequate by the planning authority (Cowley 1967). For this reason the official record of dereliction in the brickfield (165ha) is much lower than the total amount of derelict land, including that covered by unsuitable or unenforceable planning conditions (approximately 560ha). The brickfield has long been a source of contention between the manufacturers and the planning authority, and only in the 1970s has the prospect of more than minimal restoration begun to emerge (see pp 244–5). It is estimated that the brick-clay excavations will consume a further 1,118ha during the period

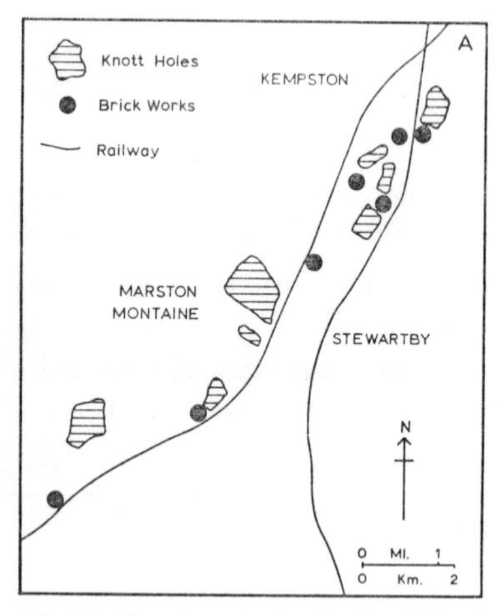

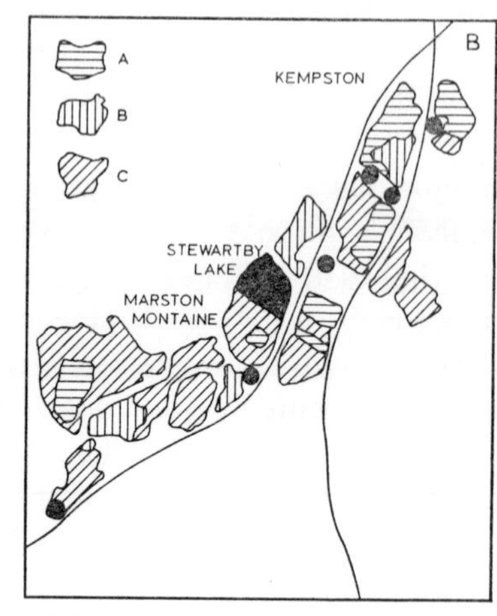

Fig 27 Land use in the Bedfordshire brickfield: (a) in 1939, showing brickworks and clay pits (knottholes); (b) in 1972
A Active pits (some partially waterlogged)
B Pits infilled or otherwise reclaimed
C Land covered by planning permission to work clay
The land in the far south-west is unlikely to be worked. Data from Bedfordshire County Planning Office

1972–2033, but as planning conditions are much more stringent than in the 1940s, it is unlikely that long-term dereliction will result from this working.

(*ii*) *China Clay in Cornwall.* The china clay deposits of Great Britain are as concentrated as the various brick clays are widespread, and it is this high degree of concentration which contributes to the distinctiveness of the landscape created by china clay working. The major deposits are in the St Austell district of Cornwall, with subsidiary outcrops on Bodmin Moor and on the southern margins of Dartmoor in Devon. China clay is formed in the natural decomposition of felspar, one of the principal constituents of granite. In Cornwall and Devon the mineral decomposed *in situ* and the china clay was formed by hot solutions ascending from beneath the earth's crust, thereby forming virtually bottomless deposits. In addition the worked deposits are free from mineral traces that might stain the clay and render it unmarketable. It should be noted, however, that within the clay-working districts there are islands of partially kaolinised granite, and in some localities the clays have been discoloured and rendered worthless. The worked deposits are homogeneous and, even where the maximum depths of working have been reached at over 115m, the china clay is still of very high quality. In its natural form china clay contains rock fragments and fine-grained detritus which must be removed before the clay is saleable. As china clay has a low value in relation to its bulk, this waste must be removed near the pits in order to minimise transport costs.

There are, therefore, two characteristics of china clay working which may give rise to dereliction: one is the depth of the excavations, the other is the large amount of waste which has to be dumped close to them. In addition the concentration of the workings in a relatively small area (the major deposits near St Austell, where eighty-five pits were at work in the 1950s, cover about 35km^2) has produced difficulties in siting the dumps satisfactorily in relation to further expansion of the pits. Even though it may be possible to increase output by working to depths 30m below the present pit bottoms, it is necessary for this deepening to be accompanied by lateral expansion in order to grade the sloping sides properly. A further source of dereliction has been the adoption of new methods of processing the clay, which has led in recent years to the abandonment of drying kilns and the mineral railways that used to serve them. In many instances the old clay pits have become permanently flooded, but not necessarily derelict, as they may function as reservoirs for the water used in sluicing the clay at active pits. After sluicing, the coarse materials are separated from the liquid clay and dumped on the adjacent waste mounds. Further refining takes place at settling ponds before the clay is finally prepared for sale by filtering and drying. The ratio of waste to marketable clay is approximately 8 : 1, excluding the overburden removed from the pits. This is rarely more than 8m thick, and constitutes a tiny proportion of the total yield of waste.

The landscapes produced by china clay working (see plate, p 50 and Fig 26) are dominated by deep pits, some of them permanently flooded, and waste heaps (locally term 'burrows'): those in the right foreground of the photograph illustrate the severity of gullying which may occur and also show that thin scrub-like vegetation is able to establish itself on the untreated tips. Opinions differ as to the ugliness of this landscape and the extent to which dereliction is a major land-use problem in the locality. S. H. Beaver (1969) observed that the turquoise blue lakes formed in the abandoned clay pits 'would undoubtedly prove a great attraction if [they] were natural and not man-made!' The waste heaps are distinctive features in an otherwise unremarkable landscape, and they mainly sterilise low-grade land. Perhaps the major problems stem from the complex intermingling of active and disused pits fringed by waste heaps of varying size, which largely reflects the haphazard evolution of the industry (Barton 1966). This militates against any form of phased reclamation of worked-out areas, and further complications ensue from the intermixture of workings and mining settlements.

Consequently the china clay district of St Austell in particular poses severe problems of mineral extraction and land use, and these are rendered all the more intractable by the fact that the waste materials lie too far from major markets for fill or for building materials that might be made from them. This difficulty might, however, become less important if costs of working traditional sources of aggregate continue to increase, or if local authorities in areas such as South East England adopt more stringent planning policies towards sand and gravel working.

(*b*) *Deep workings in 'hard' rock*

(*i*) *Slate quarrying in North Wales*. The Welsh slate quarries provide a particularly daunting example of dereliction produced by quarrying, partly because of the location of the workings in areas of considerable amenity value, partly because of the scale of the excavations and the accompanying waste dumps. It may be argued that both the terraced quarries and the scree-like dumps are more readily assimilated by their strong mountain landscape settings than would otherwise be the case. But the fact remains that derelict slate workings constitute an important planning problem in several localities lying close to or within the Snowdonia National Park (Jacobs 1972).

The slates are found in various metamorphic rocks which display slaty cleavage, mainly in Cambrian and Ordovician formations and to a lesser extent in the Silurian. The principal workings for Cambrian slates are situated on the narrow outcrop skirting the western edge of Snowdonia for a distance of 18km or so between Bethesda in the north and the Nantlle valley in the south. Large-scale quarrying began in the 1780s, and during the nineteenth century the Penrhyn and Dinorwic quarries emerged as the largest. The Penrhyn Quarry comprises 75ha of terraced working faces flanked by over 200ha of slate waste (Fig 26). The Dinorwic Quarry, closed in 1970, covers almost 300ha of mountainside, with a difference in altitude of 550m between the highest and the lowest quarry faces. In the Nantlle valley the slates were formerly worked at a number of deep quarries which, unlike the terraced workings further north, were liable to flood and collapse of their steep sides. This locality contained approximately 340ha of derelict slate workings in the late 1960s. The Ordovician slates of the Blaenau Ffestiniog district contain the remaining major workings. Here the slates are at a gentle angle of dip and they were extensively mined, although in several instances the underground galleries were opened up so that the former supporting pillars could be quarried. Thus, the principal land-use problem produced by the Blaenau Ffestiniog workings is the disposal of waste slate.

The volume of slate waste produced at the Welsh quarries averages 20 tons for every ton of saleable slate and arises from the highly selective process of slate dressing. At most quarries little or no overburden has had to be removed, except in the Nantlle valley. The visual impact of the waste dumps aroused unfavourable comment when the boundaries of the Snowdonia National Park were being determined in 1947, and as a result all the major slate workings were placed outside the limits of the park (including those at Blaenau Ffestiniog, which form an enclave within its territory). This administrative device did nothing to solve the problems of dereliction, which have become worse with the progressive closure of the quarries. The prospect for reclamation on conventional economic grounds is slight and any improvement will have to be cosmetic, which is appropriate to an area of great scenic value. C. T. Crompton (1967) has indicated the possibilities for establishing vegetation on slate waste, but the scale of the problem is great. Very little has yet been done to reclaim the slate workings, which probably account for about 1,000ha of derelict land in Wales. Onyl oen major scheme, the Padarn Country Park, Llanberis, is as yet likely to impinge on the areas of derelict slate quarries and waste dumps.

(*ii*) *Limestone and chalk quarrying in the United Kingdom.* Of the many forms of non-metalliferous mineral worked at quarries in the United Kingdom, limestone is the most important in terms of volume of production (Fig 6). During the 1960s output more than doubled (from 41 million tons to 88 million tons, including a little from deep mines), although the number of active quarries was reduced by 20 per cent. In the nature of things limestone quarries are often situated in areas of great amenity value, and this is particularly true of those in the Peak District, the Mendip Hills and on the coast of north Antrim. Parts of the Carboniferous Limestone outcrop in the Peak District have long been worked as a source of lime for agricultural and industrial use (Boden 1963), and here, as in Snowdonia, the boundaries of the National Park were drawn to exclude those localities with the greatest intensity of quarrying in the 1940s. The principal exclusion covered Dove Holes Dale east of Buxton where, by the 1960s, there was extensive dereliction

Page 175 Derelict land at Winsford, Cheshire, 1964, showing the sites of saltworks abandoned at various dates between 1920 and 1954: the tracks of disused mineral railways can also be seen (see also plate, p 85)

Page 176 Derelict land in the Black Country, Staffordshire, 1971. For an interpretation see Fig 42

comprising abandoned quarries and their waste heaps, together with disused kilns, silos and railway sidings. In the Mendips limestone quarrying has removed entire hills and lowered the skyline by as much as 30m in places (Hall 1971). In north Antrim derelict quarries and ancillary plant are particularly noticeable at Carnlough, but there are several other points along the scenic coast road between Larne and Portrush at which abandoned or active quarries dominate the view (Wallwork 1973). The most intrusive quarries are those which have been carved out of a hillside or a cliff face, but these are not necessarily the most problematical as derelict features. Deep quarries, which under modern methods of working may be as much as 25–30m deep, present far greater problems of reclamation than either the cliff-like excavations or the shallower workings produced in the early days of quarrying, because they may be dangerous if left unfilled, and are unlikely to be so treated in the absence of suitable filling materials (Jacobs 1973).

Part of the growth in volume of limestone output has been due to the increased demand for cement since 1945 (Grimshaw 1968). There are numerous cement works situated close to limestone quarries – for example, at Clitheroe in Lancashire and Aberthaw-Rhoose on the coast of Glamorganshire. Cement manufacture is, however, traditionally associated with chalk quarrying, and there are particularly large and long-established works on Thames-side and in the Medway valley. On Thames-side the industry is located where the river breaches the chalk outcrop, although with the passage of time most of the workings have moved a little distance from the tidewater. In Essex the quarries and cement works are in the area between Grays and Purfleet (Robinson and Wallwork 1970), and on the Kent shore they are centred on Swanscombe (Davis and Baker 1964). In both localities deep disused chalk quarries constitute a major, if highly localised, problem of land use. At its maximum the unconsolidated overburden is 30m thick, as is the pure Upper Chalk, giving an overburden to mineral ratio of 1 : 1 but producing very deep workings.

Chalk quarrying on a large scale began in the 1870s and by 1950 the Swancombe workings covered over 400ha (Coleman 1954). Two major problems have arisen from the mode of excavation. Firstly, several of the quarries have been abandoned and, although some have been infilled and others have provided sites for industrial development (eg Bowater's paper mill at Northfleet), many remain derelict. Secondly, a two-tier landscape has evolved, with communications and settlements perched above the excavations. At best this is inconvenient for everyone concerned, and at worst it has prevented the logical extraction of the chalk and the phased reclamation of abandoned workings. It is instructive to compare the experience of this locality, in which quarrying has been strongly influenced by the accidents of land ownership to the long-term disadvantage of both the industry and the communities which rely upon it, and that of the Cologne brown-coal field considered below.

(*c*) *Deep workings of fossil fuels: brown-coal or lignite*

There is no example of this type of working, with its low ratios of overburden to mineral-seam thickness, in the British Isles. In Europe, however, exceedingly thick beds of brown-coal are worked in various parts of Germany (Elkins 1953a, 1956) and in north Bohemia (Förster 1971), and there are even more impressive depths of opencast mining in the Australian state of Victoria (Scott 1958). The principal field in the German Federal Republic lies to the west of Cologne and is of interest not as an example of dereliction but as an illustration of the application of well tried and long established legal controls devised to minimise long-term disruption and damage (Elkins 1953b), The Cologne brown-coal field is largely coincident with the Ville ridge and comprises a main seam ranging from 9 to 90m in thickness, with subsidiary thick seams, all of them almost horizontal. The overburden of Tertiary and later superficial deposits, including loess in the northern part of the field, is unconsolidated, and the coal itself is friable – conditions which have encouraged intensive mechanisation. Furthermore the ratio of overburden to coal is rarely more than 1 : 1, although this means that excavations over 180m deep exist.

The mode of working these thick seams differs little in principle from any other form of opencast mining. The important difference, as evident in the landscape after mining, stems from German law. In mining law there are two basic assumptions: firstly, mineral extraction is economically vital and must therefore proceed with the least possible hindrance; secondly, in return for this privileged position the mineral operator must accept an obligation to restore the areas which have been worked. Thus the extensive deep pits which mark the Cologne brown-coal outcrop (Fig 26) have, in many instances, replaced entire villages, which have been relocated elsewhere, and caused major traffic arteries to be rerouted. On cessation of working the mines must be restored by dumping overburden from pits which are still active, ensuring that the final level is above the water table. During the 1930s, legislation was framed to ban the creation of spoil heaps other than in exceptional circumstances. Previously spoil heaps had been created and pits left waterlogged, particularly in the southern sector of the field near Brühl. However, the tips have been afforested and the flooded workings have become a valuable amenity as man-made lakes.

The Cologne brown-coal field provides an excellent example of the force of legal constraints working to the advantage of a mining district. The kind of dereliction that could have ensued is clearly evident from the massive heaps of overburden which await return to the pits in the northern sector of the field or removal by rail to fill-deficient areas further south, and by envisaging the pits themselves as massive lakes. A somewhat different pattern of land use has emerged in the north Bohemian brown-coal basin, which produces 97 per cent of Czechoslovakia's brown coal (Förster 1971). Deep mining accounted for 15 per cent of the 54 million tons produced, the remainder being worked opencast. Opencast mining had grown at a more rapid rate than deep mining in the post-war period, a fact partly reflected in the higher proportion of dereliction attributed to it.

In 1965, 38,000ha of devastated land were recorded in the region, 78 per cent having been caused by opencast mining. Set against this, the rate of reclamation was very low, increasing from 134ha per annum in the early 1960s to 500ha per annum after 1970.

(*d*) *Metalliferous minerals*

Deep opencast mining of metalliferous minerals has become increasingly important as the working of low-grade ores has gained momentum. Almost all the workings of this type lie outside Europe and are particularly evident in the mineral-producing areas of newly developing countries. The Panguna copper mine, Bougainville (Cummings 1972), provides an example of the waste-disposal problems posed by such workings. During the preparatory stages of working in 1968–72 some 300ha of rain forest and riverine swamp were covered with tailings to a depth of between 4–5m, even though about 60 per cent of the debris produced had been swept out to sea. The active mine will continue to yield tailings and will also create huge rock dumps during its 30 year lifespan, at the end of which a pit 300m deep and 170ha in extent will have been created. It may be argued that the existence of this and similar mines in tropical rain forests or arid deserts is less damaging than would be the case in, say, Snowdonia, but the fact remains that they cause great upheaval and disruption wherever they may be located. Indeed it could be claimed that, in the long term, workings such as those formerly proposed for Snowdonia and the Mawddach estuary would be less damaging than their overseas equivalents, if only because mining would be carefully controlled and restoration of the sites guaranteed. There are, however, persuasive arguments for concentrating future expansion of mining in the developing countries, irrespective of the case against loss of amenity in Great Britain (Warren 1973). But such is the demand for metalliferous ores that the opening up of deep pits in European mining localities which have long been defunct is considered to be economically worthwhile, as at Tynagh, Co Galway (Orme

1970), where workings for lead-zinc-silver ores may ultimately reach a depth of 110m.

Deep opencast mineral workings are, and long have been, commonplace in the United States. For example, deep pits were being worked for haematite in the Lake Superior orefields at the turn of the century, according to the annual reports of the US Geological Survey. By the 1940s the opencast mines had reached depths of 120m and the largest single working covered an area of 500ha. The Lake Superior mines were given a new lease of life in the 1950s by the discovery that taconite, a low-grade iron ore, was commercially usable. It is considered possible that on abandonment the Mesabi Range iron ore pits will form a single steep-sided lake 190km long and 5km wide, fringed by extensive spoil banks (USDI 1967). The trend towards concentration of lean ores before shipment from the mining areas has increased the volume of waste dumped in these localities. Thus 3 tons of taconite has to be worked and concentrated in order to produce 1 ton of iron ore with a 67 per cent ferrous content (Kohn & Specht, 1958). Between 1955 and 1965 21 million tons of waste tailings from the concentration plants were dumped into Lake Superior, causing excessive emission of pollutants. Similar problems stem from other forms of low-grade ore extraction. Thus at Bingham Canyon, Utah, a massive opencast copper mine is effectively removing a 470m high mountain of low-grade ore. As at the taconite mines, the concentration process gives rise to large volumes of waste, with the added danger that the dumps themselves may ultimately be reworked by methods which cause excessive pollution. It can be argued that many of these large opencast mines producing low-grade ores are located in areas which have no other economic bases, and that no land-use conflicts arise. It would certainly be impossible to reclaim the workings by infilling, and in fact this is never likely to be thought necessary. The major difficulties stemming from their existence are much more likely to be problems of pollution rather than those of dereliction, as commonly understood in Europe.

TYPE 4: SURFACE WORKINGS BELOW THE WATER TABLE

Several minerals occur in unconsolidated deposits lying beneath the water table, and are capable of being worked either by conventional excavators or by dredging equipment. A distinction may be made between 'pure' workings, at which little or no detritus is produced, and 'residual' workings, at which the mineral is extracted from a greater volume of material, most of which is left as tailings.

(*a*) *Wet sand and gravel pits*

In Great Britain during the 1960s approximately 1,600ha of land per annum was being worked for sand and gravel. About 80 per cent of the pits were wet workings, the mineral being extracted by various means from beneath the water table. In addition, increasing amounts of sand and gravel were being got from marine dredging and this output from offshore deposits accounted for roughly 10 per cent of the 109 million tons produced in 1971. The location of sand and gravel workings is largely determined by two sets of factors. Proximity to the market is of crucial importance, as the ex-workings price per ton doubles at a distance of 32km from the pit unless bulk haulage facilities are available. Physical accessibility of the deposits is also important and it is unusual to work sand and gravel from beneath a greater thickness of overburden than 4m. Wet gravel pits are mainly located on river terrace gravels, where the overburden is normally thin but almost invariably constitutes high-quality soil. Land of this kind is commonly used for intensive arable farming or horticulture, particularly near large towns, where there is also likely to be heaviest demand for sand and gravel. Not surprisingly numerous instances of land-use conflict have emerged in gravel-bearing localities where wet working is practised, notably in the London region.

Wet sand and gravel workings produce little waste other than unsaleable residues separated at pit-side washeries, and these form a very small proportion of the total volume worked. The principal source of potential dereliction stems from the fact that on abandonment the pits are waterlogged, and this was long considered to be a major obstacle to their reclamation (Wooldridge and Beaver 1950). Even in the London region, where there was both pressure on space and large amounts of refuse to be disposed of, there were large areas of derelict sand and gravel workings in the late 1940s. Thus in Middlesex in 1947 there were 400ha of derelict sand and gravel workings along the Lea valley in the east and, more particularly, along the Colne valley in the west. The threat of accelerated dereliction in the Greater London green belt led to a proposal that sand and gravel working should be severely curtailed in that area (Abercrombie 1945). This was ruled out on economic grounds and, although more stringent planning conditions were attached to new workings after 1947, sand and gravel extraction continued in the green belt and beyond. Thus in the Colne valley there were over fifty pits at work in the late 1960s, several of them 10–15ha in extent, forming a chain of man-made lakes 8km long and almost 1km wide between Rickmansworth and Denham. Within the Colne valley 1,320ha of gravel had been worked at both wet and dry pits by 1969, and half of this area lay unreclaimed after use. In addition a further 500ha of gravel-bearing land had been released under planning control for future working. Whether wet sand and gravel pits pose the intractable problem of reclamation that was supposed in the 1950s is, however, open to doubt, as inland water is now at a premium for amenity uses (see pp 246–7).

(*b*) *Dredged mineral workings*

Several types of metalliferous ore occur as fine-grained particles within unconsolidated superficial deposits of sand, gravel and alluvium. Where these minerals lie below the water table, they may be worked by suction pumps or mechanical dredgers, in the same way as sand and gravel. The important difference is that in working metalliferous ores by these techniques a high proportion of the parent material is left as waste tailings at the point of extraction. Extraction of metalliferous ores by dredging has been uncommon in Great Britain, although proposals to dredge tin were seriously considered in 1913 at Bugle, Cornwall, and for a short period in 1925 a tin dredger was at work on Goss Moor in the same locality (Barton 1967). More recently proposals to employ suction dredges offshore in St Ives Bay have reintroduced the possibility of working tin ore along the Cornish coast, but as yet detrital tin is worked by methods which do not produce waste tailings on a large scale. Consequently it is necessary to take examples from overseas in order to illustrate this form of mineral working.

The principal minerals worked by dredging are tin and gold. Tin dredges were introduced to the Malaysian alluvial deposits in 1912, but from the 1880s onward tin has also been worked by suction pumps. The major areas of gold dredging have been located in Alaska, California and New Zealand. In Alaska the placer gold deposits have been reworked by dredgers, and the resultant destruction of the permafrost is said to have made the tailings valuable as building sites (USDI 1967). The same benefits have not, however, accrued from the New Zealand dredgings, where gold dredging machinery was first introduced to the Otago fields in 1863 (Forrest 1965) and during the 1880s steam-powered dredgers were first employed. At the turn of the century over 100 powered dredgers were working the goldfields of Otago, Southland and Westland (Cumberland and Whitelaw 1970). As they moved along the alluvial flats and gravel terraces, they left behind a trail of waste tailings similar in form to the hill and dale of opencast mining. The sub-humid climate of Otago has ensured that the tailings remain raw and unvegetated long after working has ceased. In the humid valleys of Westland, where the last dredger was at work near Taramakau in the 1970s, extensive scrub-covered tracts of tailings are a common feature, having in many instances destroyed what little farmland there was on the narrow coastal plain.

CONCLUSION

During the twentieth century surface mineral working has become the most disruptive form of extractive industry, for it has greatly increased its output at a time when deep mining has diminished. As a rough approximation the combined output of surface workings in the United Kingdom has increased by 500 per cent since 1900, whereas that from deep mines has fallen by 45 per cent, though these figures give no more than a very general impression of the relative importance of the two types of working. In some instances surface working has invaded localities that were once the preserve of deep mining, as, for example, in the Appalachian coalfields. In others the development of surface working has adversely affected deep mining of the same mineral in distant localities: for example, the rise of the hydraulic tin mining industry in Malaysia after 1880 was a major contributory factor to the demise of Cornish tin mining.

Surface mineral working is disruptive for three main reasons. Firstly, it commands exclusive use of the mineral-bearing land, unlike deep mining, which permits other activities to be maintained above ground in spite of the risk of subsidence. Secondly, the process of excavation inevitably causes upheaval, through the stripping of overburden and the removal of the mineral deposits. Shallow workings are not necessarily less disruptive than deep ones, for they may cover much greater areas and, if left waterlogged on abandonment, may restrict the range of options for reclamation; but they are at least normally capable of being reclaimed, unlike many of the deeper workings. Thirdly, spoil heaps are as likely to be a feature of surface workings as of deep mines. They may contain overburden, interbedded barren measures and waste material which forms a constituent of the mineral being worked. Thus at an opencast coal mine overburden and unproductive interbedded deposits are likely to form the bulk of the waste, with debris from the coal an additional element if there are washeries at the mine. The waste at a china clay pit on the other hand largely comprises sand and rock fragments from the mineral, with very little overburden or barren material.

The dereliction caused by surface mineral workings is a product of the abandonment of both excavations and waste heaps, coupled with the survival of disused plant and ancillaries such as screens, washeries, silos, kilns and transport installations. In many cases it is the ruins of plant and buildings which give greatest visual offence at abandoned surface workings. A cliff-like quarry face can readily be assimilated by the surrounding landscape; rusting hoppers and corrugated iron sheds or the crumbling masonry of kilns cannot. Although surface mineral workings cannot avoid being disruptive, unlike deep mines which can stow waste below ground, it has proved possible to substitute immediate reclamation of land for preventive measures in many instances. Reclamation is properly the concern of later chapters, but it is important to recognise here its significance when employed as part of the process of extraction. The open-cast coal mines and ironstone quarries of Great Britain and the brown-coal pits of West Germany provide examples of forms of surface working which, however disruptive during the period of active extraction, have caused no long-term dereliction.

References to this chapter begin on p 313.

CHAPTER SIX

Industrial Dereliction

WHEREAS mining and quarrying normally produce dereliction of a predictable form and extent, it is more difficult to generalise about the types of derelict land which stem from other classes of economic activity. For reasons which are explained below the examples of industrial dereliction discussed in this chapter are largely confined to those produced by mineral-based industries. In addition, certain service industries are also considered, notably transport and public utilities, as several of the forms of dereliction associated with them have produced problems not normally encountered in either mining or manufacturing. All the examples are drawn from within the United Kingdom, but the forms of dereliction are universal and could equally well be illustrated from other parts of the world.

There is a reasonable certainty that mineral extraction will ultimately cause dereliction unless positive steps are taken to prevent it, but as industrial processes move further from the stage of mineral extraction, both the predictability and the likelihood of dereliction diminishes. Thus mineral-based industries may produce varying amounts of solid residues as they refine and process mineral raw materials. If these industries are tied to local sources of raw materials, their exhaustion may lead to the abandonment of the processing plants as well as the mines and quarries which served them. This is not invariably so, for many mineral-based industries have survived the exhaustion of local reserves by importing raw materials, thereby avoiding closure or deferring it. Nevertheless it is reasonable to suppose that mineral-based industries will create derelict land at some stage in their existence, even though this is less predictable than it is in the case of mineral extraction.

Manufacturing industries which use non-mineral raw materials, or which are based on the fabrication of semi-finished goods or manufactured components, are much less likely to produce dereliction through waste disposal because their residues can be recycled or disposed of by controlled tipping and other means normally applied to the treatment of towns' refuse. This is not to disguise the fact that some manufacturing industries of this type do create dereliction by illicit or uncontrolled tipping, often on land which is already in poor physical shape, but in principle it is impossible to generalise about such practices and the related problems of illegal disposal of trade effluents. Manufacturing industries may also create dereliction through the closure of factories and the abandonment of transport installations such as private railway sidings and canal basins, but in ways which are again difficult to generalise about. Many of the buildings abandoned owing to the contraction of manufacturing industry have proved to be capable of adaptation to other uses at little financial cost, in a way that has never been possible with, say, a derelict colliery or ironworks. For example, in the North West Region roughly 4,950,000m^2 of factory space was vacated during the period 1967–70, but at the same time an area equivalent to 85 per cent of this floor space was reoccupied. Even so 711 industrial premises lay empty in the North West at the end of 1970, one-quarter of them textile mills, and although it is obvious on the basis of past experience that many will be reoccupied, a small proportion may become derelict.

MINERAL-BASED INDUSTRIES AND DERELICTION

Industries which refine or otherwise process minerals are in varying degree potential sources of derelict land, principally through the need to dispose of solid wastes. Although many minerals are partly refined or concentrated at the point of extraction, this is not universally so, and it was less common in the past than at the present day. Solid wastes are normally composed of impurities in the various raw materials and fuels employed, or, less commonly, include material introduced to facilitate the manufacturing process but which does not form part of the final range of saleable products. The classic example of waste combining both characteristics was provided by *galligu*, a noxious residue of the Le Blanc process. It contained impurities from the salt, limestone and coal used in the process, together with calcium sulphide which incorporated the sulphur introduced as a catalyst at great expense (one-third of the cost of raw materials, but only one-tenth of their weight). In addition the process also liberated hydrochloric acid gas into the atmosphere. Although techniques of preventing this emission and of sulphur recovery were eventually adopted, substantial dumps of solid waste from factories which closed over 60 years ago survive on Tyneside and in the Widnes-St Helens district of Lancashire (Fig 28).

The constituents of waste dumps produced by mineral-based industries may conveniently be divided into two types: slags from the smelting of metalliferous ores, and slimes derived largely from processing non-metalliferous minerals. Slag is most commonly associated with iron and steel manufacture and slag tips were normally found at every iron and steel works. The composition of slag and the quantity produced in relation to output of iron and steel varies considerably with the chemical constituents and ferrous content of the ore and the form of smelting employed. Where very lean ores are employed, the ratio may be high: thus at Scunthorpe 1·3 tons of slag are produced

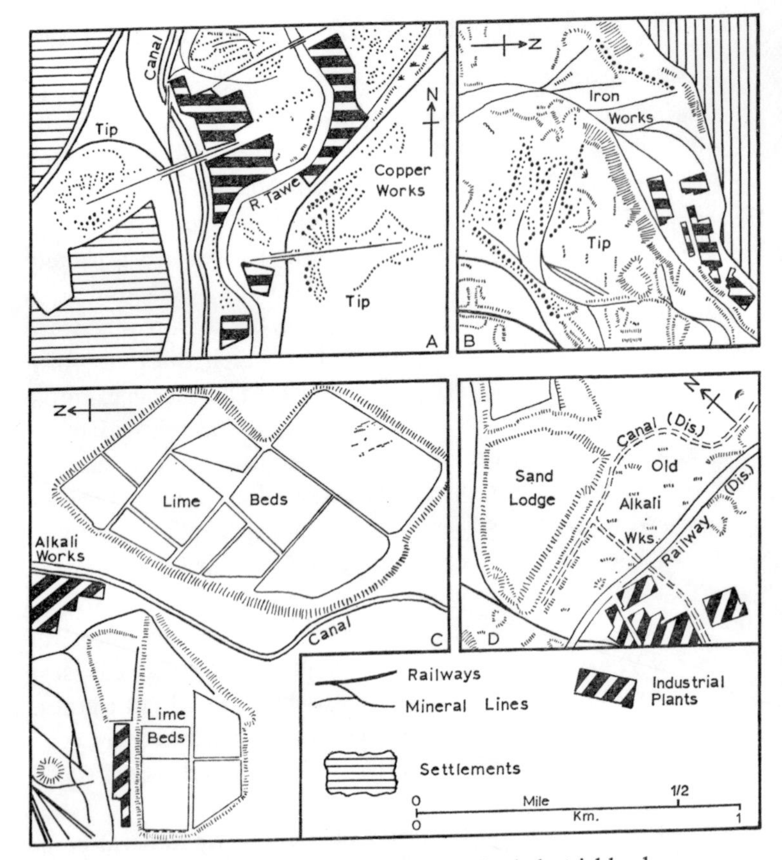

Fig 28 Miscellaneous forms of potentially derelict industrial land

- A The Swansea Valley, c 1920, showing copper smelting works and associated tips
- B Dowlais Ironworks and the Great Tip, c 1920
- C Waste dumps ('lime beds') at alkali works, Lostock Gralam, Cheshire: see plate, p 158, and text, pp 192–3
- D 'Sand lodges' at glassworks, and disused Le Blanc alkali works' sites, St Helens, Lancashire. Two other common elements of dereliction – disused canals and railways - are also depicted.

All the maps are to the same scale and industrial plants generalised

for every ton of pig iron, and this ratio is said to be the least favourable in the world (Pocock 1963). Fortunately it has proved possible to develop markets for slag and, through controlled smelting, to obtain a quality-controlled by-product. Since the 1920s slag has been commonly used as a raw material for the manufacture of tarred roadstone and as a constituent of other constructional materials. As long ago as 1887 basic slag from steel furnaces was being used as an agricultural fertiliser, and since 1945 output of basic slag (expressed as phosphate content) has averaged 80,000 tons per annum. In addition to using slag direct from the furnaces many of the once derelict slag heaps have also been worked commercially. However, many tips remain in which the slag has a high sulphur and lime content or a honeycomb structure, rendering it less useful as hard core. Particular difficulties surround the use of tips produced by the smelting of acidic haematite ores, as in West Cumberland and along the north crop of the South Wales Coalfield. For example, the Dowlais Great Tip (Fig 28) covering 60ha and abandoned when the iron and steel works closed in 1930, has only recently been worked, and the reclamation of its site has been delayed by the difficulty of finding suitable commercial outlets for the grey fused slag of which it is composed.

Other forms of smelting crude ore also yielded slag, even where concentration had taken place at the mine. Particularly striking examples were to be found in the lower Swansea Valley (Fig 28). Between 1717 and 1850 thirteen copper smelting mills were established in this locality using the 'Welsh process', which involved calcining the ore in order to separate the copper from the iron, silica and other solid residues that were dumped on the slag heaps and ash tips. After the demise of the industry in the lower Swansea Valley in 1928, the solidly built plants and massive slag heaps were to prove particularly intractable forms of dereliction (Hilton 1967), as were the waste tips of lead and zinc smelters in the same locality. Elsewhere in Great Britain copper smelting produced pockets of derelict land at Amlwch

(Anglesey), Cheadle (Staffordshire) and St Helens. Waste from lead smelt mills was found on a small scale at numerous places on the Pennine orefields, in Flintshire and in the Mendip Hills. At many smelt mills the slag heaps were reworked from time to time in order to recover various minerals. As late as 1950 the Nenthead lead smelter in the north Pennines was re-equipped to extract lead, zinc and fluorspar from the spoil dumps (Raistrick & Jennings 1965).

Slimes produced by mineral-based industries present greater problems of potential dereliction than slag, partly because they require the construction of space-consuming reservoirs or lagoons for their disposal, and partly because the waste may be exceedingly difficult to reclaim when it solidifies. At best it may only have the load-bearing qualities of unconsolidated alluvium: at worst it may never be capable of more than simple cosmetic treatment to reduce its visual impact. Similar problems are also experienced with slurry ponds at coal washeries, except that they are frequently worked for coal waste and in most cases form part of complexes of potential dereliction rather than occurring as separately identifiable features. Several mineral-based industries include settling and slurry ponds among their installations, notably cement, lime, salt and glass works (in the last-named instance termed 'sand lodges', as depicted in Fig 28).

The most distinctive and potentially most damaging slime deposits are those produced at ammonia-soda works, of which there are numerous examples on the Cheshire saltfield (see pp 272–5 and Fig 28). Termed 'lime beds', the waste reservoirs contain impurities from the raw materials used in the ammonia-soda process, together with calcium chloride, which ultimately drains away, leaving semi-deliquescent solids. Until the 1950s the waste was dumped on the surface, either on land already damaged by salt working subsidence or, increasingly, on sound agricultural land. The prospect of having to consume even larger tracts of good farmland caused the industry to investigate alternative means of waste disposal, which led to the adoption of a method of underground stowage during the early 1950s

(Wallwork 1960, 1973). This did not, however, remove the backlog of previous dereliction and, for this reason, mid-Cheshire still contains several problem areas in which lime beds are a prominent feature of the derelict landscape.

The closure of plants engaged in mineral-based industrial processes almost invariably creates more severe dereliction than with other factory closures. Firstly, mineral-based industries commonly employ specialised buildings and installations which are not easily adapted to other uses as they stand. Abandoned offices, workshops and warehouses at the plant may find new occupants, but the specialised installations are far less likely to. Secondly, the removal of those installations which have scrap value normally takes place quickly but leaves behind sites that are expensive to reclaim for other purposes. The massive foundations of blast furnaces, calciners and chemical plant, together with the ducts and pipe-trenches which intersect the sites, make clearance costly. Thirdly, the fact that many of the abandoned works are surrounded by dumps of waste or disused mineral workings diminishes their attractiveness to new occupants unless finance exists to initiate major schemes of reclamation. Fourthly, it was not uncommon for such sites to be bound by restrictive covenants which limited their use on abandonment, or continued to place conditions on the right to abstract water and minerals. Consequently many mineral-based industrial sites lie derelict long after their closure. For example, the sites of the Sandbach and Plumley ammonia-soda works, Cheshire, closed between 1925 and 1930, had not been reclaimed over 40 years later. The Cyfarthfa ironworks, Merthyr Tydfil, abandoned in 1924 (see plate, p 68) lay derelict until 1938 and its slag heaps remain. Similarly, many of the abandoned smelting works in the lower Swansea valley were not cleared until more than 40 years after their closure.

PUBLIC UTILITIES AND DERELICTION

Public utilities provide variant examples of industrial dereliction which in principle resemble those attributable to the manufacturing sector. Dereliction falls into two categories: abandoned buildings, and spoil dumps. Abandonment of buildings and plant has been particularly widespread in the electricity generation and gas-producing industries since 1945, mainly because of technical developments in both manufacture and transmission. Thus there was a net decrease of electricity generating stations from 297 in 1949 to 183 in 1972, and of gas-works from 1,050 in 1949 to 96 in 1972. As both industries had developed through the establishment of numerous market-orientated plants, almost all the closures were of obsolete town works. (In 1949 75 per cent of generating stations had an installed capacity of less than 100,000kw, and 50 per cent of gasworks produced less than 500,000 therms per annum.) Few of the sites remained derelict for long, either being maintained as storage stations and depots by the industry concerned or attracting new forms of development in relatively valuable urban locations. Consequently neither derelict generating stations for gasworks feature prominently in the lists of grant-aided reclamation schemes produced by local authorities. Sewage disposal plants have also been subject to schemes of rationalisation since 1945, mainly through the closure of small obsolete works and their replacement by large centralised installations.

The spoil produced by public utilities comes from two major sources – sludge beds at sewage works and furnaces at electricity generating stations. Sludge from sewage works is treated in a variety of ways, much of it ultimately beneficial in that it can be composted or used rawin reclamation schemes. In some instances the treated sludge is dumped at sea, as, for example, from the Greater London treatment plants at Beckton and Crossness, which dump over 3 million tons of waste per annum in the Black Deep off the Thames Estuary. Problems of disposal exist at those works where the treated sludge contains toxic industrial wastes, and this causes dereliction when the residues have to be dumped on land.

The disposal of furnace ash from electricity generating stations first became a major land-use problem after 1945, when production began to be concentrated at a smaller number of large stations and the use of pulverised coal with a high ash content greatly increased. This produced a very fine dusty ash termed pulverised fuel ash (PFA), and its disposal threatened to cause considerable difficulties for the electricity industry. In 1949 the British Electricity Authority's annual report suggested that the increased volume of solid waste would have to be dumped in abandoned surface mineral workings, on low-lying land or, exceptionally, at sea. By 1957 ash tipping was officially considered to be 'a serious and increasing disposal problem', even though PFA (including furnace bottom ash) only accounted for half the 6·8 million tons of waste produced at generating stations. Two years later the proportion had risen to 70 per cent of a slightly increased yield of furnace waste, but by this time the Central Electricity Generating Board (CEGB) had begun to investigate possible commercial uses for PFA, as well as examining means of reclaiming existing dumps. During the 1950s the CEGB began commercial manufacture of building blocks and aggregate from PFA and also marketed raw ash as a filling material for road construction. Between 1958–9 and 1971–2 the percentage of PFA sold commercially had risen from 6 per cent to 57 per cent of output at a time when production of ash had risen from 5·2 million tons to 9·4 million tons. Indeed by 1972 supplies of PFA in south east England were inadequate to meet demand, partly because of greater efficiency in coal burning but largely due to the success of marketing policy. Thus PFA, far from being a major source of potential dereliction, had become a raw material of the construction industry in its own right, as well as being an important source of fill for reclamation schemes (pp 247–8).

DEFENCE INSTALLATIONS AND DERELICTION

Abandoned defence installations share many of the characteristics of both derelict factories and disused elements of the transport system. Land under military ownership is frequently viewed in peacetime as a particularly offensive source of dereliction. The fiercest criticism is directed at training areas and ranges, though strictly speaking these are not derelict, since they are still used. Much of the dereliction caused by the abandonment of defence installations dates from 1946 onwards, for many sites have been declared redundant owing to changes in defence policy during the post-war years. Derelict land in former military-service ownership has always been excluded from the official returns of dereliction where it was covered by the terms of the Requisitioned Land and War Works Act, 1945, but this omission is to be rectified in the new survey to be introduced in 1974. Normally the Act provided that de-requisitioned sites could either be cleared of buildings by the military authorities or the owner compensated on the basis of site value alone. As the latter course was generally the cheaper and more expedient, it was frequently adopted, so that derelict buildings and other structures were left in place (Dower 1959).

The two principal types of dereliction comprise military airfields and hutted camps. Airfields have presented particularly severe problems because of their size, the existence of concrete runways and perimeter tracks, the presence of hangars and hutments, and their mainly rural locations. In 1945 there were 720 military airfields in the United Kingdom (Blake 1969); by 1965, 510 had been abandoned and further closures have taken place since then. The amount of land covered by fields abandoned before 1965 was 85,000ha, of which about 12,140ha comprised concrete runways and various types of building. Roughly three-quarters of the abandoned area had been returned to farming and a further one-tenth put to other uses, leaving about 12,500ha of dereliction. The new uses were wide-ranging, including motor racing (Silverstone), vehicle testing (Hinckley) and, appropriately, civil commercial aviation (Castle Donington and Aldergrove). Setting aside the fact that some of the new uses were incompatible with their rural settings (Miller

1973), the scale of disruption caused by the abandonment of military airfields has been much slighter than was once feared. Some of the sections of concrete runway have been quarried for hard core (Newton 1960), as well as being used as access roads and hard standing for farm machinery, so that the failure to restore all land to crops or grass has not necessarily been a total loss to agriculture. Unfortunately some of the worst derelict sites lie in or near areas of great scenic attraction, as, for example, on the coastal platform in South Pembrokeshire (Dower 1959).

There are no statistics as to the number and extent of abandoned hutted camps, most of which were built during World War II for military and related defence needs, including the accommodation of civilians drafted to work in munitions factories or on the land. In many instances peacetime uses were found for the camps as they stood, or their sites were cleared and put to other uses. Thus a group of camps built to serve the Swynnerton Royal Ordnance Factory in Staffordshire was converted to house the Post Office engineering school and its residential accommodation, a college of education and an open prison. As with abandoned airfields a major difficulty in securing new uses for some of the derelict camps centred on their rural remoteness, and derelict encampments for which no alternative uses could be found began to figure in local authority reclamation programmes during the 1960s.

TRANSPORT INSTALLATIONS AND DERELICTION

The final category of miscellaneous forms of dereliction embraces land abandoned by various forms of transport. Again there is no clearly predictable sequence of events such as can normally be established in the case of mineral workings. Nevertheless there is a substantial amount of derelict land created by the reshaping of the transport system in the United Kingdom, and much of it gives rise to peculiar problems of reclamation by virtue of its location and form.

Railway dereliction

The problems presented by derelict railway land first attracted attention when the programme of reshaping the railway system gained momentum after 1963. The abandonment of sections of the railway network was not, however, an entirely new phenomenon, although the scale of closures over a relatively short period was. From the earliest years of railway construction parts of the network had been abandoned for a variety of operational and economic reasons. Before 1923 most abandonments were made to accommodate new commercial and technical requirements, and the effects of these closures can still be seen in the landscape. Realignments for technical reasons include the Saltash–St Germans deviation of 1908 (Patmore 1966) and the new high-level approaches south of Warrington made necessary by the construction of the Manchester Ship Canal in the early 1890s. During the inter-war years many sections of branch railway were abandoned in the economically depressed areas and the practice of closing intermediate stations on lines open to traffic also began to create some small-scale dereliction. The nationalisation of the railway system in 1948 initially had two contrasting influences on the pattern of closure. Firstly, many unremunerative lines were kept in being for reasons of social policy; secondly, many sections of the network which had been created by competitive private companies lost their raison d'etre.

Table 9 shows that between 1948 and 1970 approximately 35,000km of track (expressed as single-track length) was closed to traffic. Roughly 40 per cent of this was accounted for by sidings, which comprised only 30 per cent of the trackage in 1948. This reflected the twin policies of withdrawing freight facilities from many intermediate stations, and concentrating freight-handling installations in those urban areas which had formerly been served by competing companies. The process is well illustrated in the Sheffield area, where the thirty-nine freight yards and depots which existed before 1965 were reduced to eleven as part of a scheme of modernisation (Appleton

TABLE 9 UK changes in railway land-use requirements: reduction in track length and stations, 1948–70

	1948–54	*1955–9*	*1960–64*	*1965–70*	*1948–70*	*Percentage of 1948 track length remaining in 1970*
Net reduction of track (km)						
(a) Great Britain						
running lines	805	1,737	7,840	9,875	20,257	65
sidings	322	1,979	4,683	6,243	13,227	48
(b) Northern Ireland*						
running lines	225	870	6	161	1,262	29
sidings	32	129	16	80	257	18
UK TOTAL	1,384	4,715	12,545	16,359	35,003	59
Net closure of stations (Great Britain only)						*Percentage of 1948 stations remaining in 1970*
Passenger/freight	1,089	537	1,691	740	4,057	39
Freight only	725†	81†	1,397	687	1,278	19
TOTAL	364	456	3,088	1,427	5,335	35

Data from *Annual Abstract of Statistics:*

* Before 1958 the statistics include that part of the GNR(I) system in the Republic of Ireland: this probably amounted to about 250km of single track equivalent, which was transferred to CIE on 1.10.58

† Increase, mainly due to reclassification of former passenger/freight stations

1967). The policy of urban concentration ran counter to an earlier one of establishing large marshalling yards at nodes in the railway network that were often remote from the major sources of originating traffic: thus yards such as that at Kingmoor, Carlisle, built by the nationalised railways, have never been fully operational.

In addition to rationalising freight yards and motive power depots the railway authorities also severely pruned the passenger system after 1948 (Patmore 1966), and would have done so even more vigorously had governmental social policy permitted. Not only were complete sections of the network closed to passenger traffic, the closure of intermediate stations was also accelerated. The withdrawal of passenger services also brought about the abandonment of major city stations, whose sites became battle grounds between speculative property developers and local authorities and whose buildings became derelict eyesores. The disused Manchester Central Station, with its impressive approach viaducts, provides an example of the scale of railway dereliction in urban areas created by the closures of the 1960s. Although no precise figures exist as to the amount of derelict railway land, British Rail estimates that about 2·5ha of land is needed for every kilometre of single track formation, so that a net area of about 87,500ha has been abandoned by the state railways since 1948. This is a very tentative figure and does not represent the cumulative total of derelict land in being by 1970, but it does illustrate the scale of probable railway dereliction, excluding that caused by the abandonment of privately operated track, notably in the coalfields and various ports.

Derelict railway land falls into two major categories. Firstly, the formation, which includes the track bed and engineering works left behind after movable assets have been taken away. When a section of the railway network is formally abandoned, maintenance is pared to the minimum consistent with legal responsibility. Rank vegetation rapidly asserts itself once the weed-killing trains cease to run, slumping may affect cuttings and embankments, and masonry and metal engineering structures may begin to deteriorate beyond the point of safety. Secondly, there are abandoned buildings, including stations, warehouses and engine sheds. As with any other abandoned buildings their location and adaptability largely decides whether they are converted to other uses or become derelict.

Railway dereliction poses some problems which are not common to other types of derelict land. The parcels of land are attenuated and their physical characteristics may vary considerably over short distances through the presence of embankments, cuttings, tunnels and viaducts (Appleton 1970). There are also legal difficulties surrounding the use of derelict railway land for other purposes, including reversion clauses in the original leases and legal obligations relating to the maintenance of drainage, fences and bridges. Thus some railway land may remain derelict because its physical characteristics and legal status render it incapable of other forms of use, whereas other elements in the abandoned network are readily assimilable into the surrounding farm land, or become sites for housing and commercial development in urban areas.

Canal dereliction

Derelict canals have several characteristics in common with derelict railways, notably the linearity of dereliction and the rapidity of its onset once routine maintenance has ceased. Derelict canals are not a recent addition to the landscape, for, as with the railways, some elements were abandoned when the canal system in general was still expanding. The canals of Great Britain, excluding private branch canals and basins, and giving approximate lengths, comprise the following four major categories:

1 the commercial waterways (850km);
2 the cruising waterways (1,770km);
3 the remainder waterways (965km);
4 the abandoned waterways (1800km).

Of these, only the first category has been largely free from the threat of dereliction.

The abandoned waterways, those identified as 'lost' canals by R. Russell (1971), are by no means derelict along their entire lengths. Some of the canals were abandoned long ago and their courses put to other uses. Others are now derelict in whole or in part, including the 350km or so which might be restored and added to the cruiseway system. A high proportion of the lost canals are located in rural areas, notably in southern England (560km) and the Welsh marches (112km). Surprisingly few of the lost canals are situated in areas of heavy industry: thus Greater Manchester contains 120km, and Greater Birmingham 80km, but in both localities parts of the canal system have been preserved as cruiseways, and sections of lost canal are capable of being restored to cruising standards.

The greater part of the inland waterway system is controlled by the British Waterways Board (BWB), which produced a report in 1965 suggesting that at least 2,000km of the canal network could not be sustained by commercial traffic. The whole of this length of commercially useless water would not have fallen derelict had it been relinquished by the BWB, particularly where the canals were used as sources of water supply, but there can be little doubt that much derelict land would have been created by the abandonment of the non-commercial waterways. This was avoided largely because the BWB recognised the potential amenity value of the canals and established the cruiseway system. In addition the creation of a category of remainder canals, to include those waterways which would otherwise have been abandoned almost immediately, gave an opportunity to consider the future of those sections of the canal system which might be added to the cruiseway network after suitable remedial works had been carried out. No statistics exist as to the extent of canal dereliction, but it is important to note that the amount would have been infinitely greater had the decision not been taken to maintain a large part of the network on amenity grounds without any certainty that the cruiseway system would ultimately become self-financing. Significantly, the reclamation potential for both railway and canal dereliction has greatly increased since it was recognised that the amenity use of derelict land is often as important as more narrowly conceived economic uses.

Miscellaneous dereliction related to transport

Dereliction due to railway and canal abandonment forms the most common type associated with transport networks, but technical obsolescence and changing patterns of trade have also led to the abandonment of numerous port installations, particularly but not exclusively since the 1940s. Specialised ports handling mineral traffic have been worst affected, in some instances being closed to commercial traffic (for example, the coal ports of Penarth and Maryport) and in others handling only a vestige of their former trade (for example, Hayle, the Cornish metalliferous ore port and Workington, a point of entry for imported haematite). Even where ports have maintained a large volume of traffic, changes in its composition have led to the abandonment of specialised handling facilities, notably the coal staithes at Barry, Cardiff, Sunderland and Tyne Dock. Whether this gives rise to long-term dereliction largely depends on the extent to which the port can reorientate its trade. Thus the reduction in the number of coal staithes at Barry and their complete abandonment at Cardiff released land for other forms of port activity (Hallet & Randall 1970).

At general cargo ports abandonment of obsolete docks and quays has become a particularly important by-product of the 'container revolution' since the mid-1960s. The closure of docks in the Pool of London and Surrey Commercial Docks area has produced over 200ha of dereliction, one-third of it impounded water. Some of the land, at St Katharine's Dock, has been redeveloped and the remainder commands a high price for commercial development; its continued dereliction partly reflects indecision as to its future use: for example, the Surrey Commercial Docks have variously been considered as sites for schemes of comprehensive commercial and residential development and as the town terminal for the proposed Channel Tunnel railway (Shankland Cox 1972).

Although abandoned military airfields constitute an important element of the landscape in some localities, the closure of civil airports rendered obsolete by technical advances in aviation has rarely caused more than transient dereliction. Since 1945 several airports have been closed to regular traffic; some of them remain open to private aircraft while others have ultimately been put to alternative uses. Thus Croydon airport was developed as a housing estate, Cardiff (Pengam Moors) became an industrial site, and Belfast (Nutt's Corner) was adapted as a vehicle testing ground after having lain derelict for almost a decade.

Dereliction due to changes in the road system is scarcely

evident, for the realignment of a section of road, unlike that of railways or canals, rarely leads to the abandonment of the original formation. It is much more likely to continue to function as a service road or to be adapted to other roadside uses. The virtually complete abandonment of the passenger tramway system by the 1960s caused little dereliction because almost all the lines followed existing highways, even where reserved track was built. Consequently the abandoned formation was readily incorporated into road-widening schemes or served as the centre strip of dual carriageway roads. The nearest thing to extensive dereliction caused by the developing road system is the urban decay which occurs in localities destined to be traversed by major new highways. This planning blight may affect the quality of the built environment long before any road construction takes place, but as with other forms of urban clearance, it cannot properly be spoken of as dereliction.

Transport and dereliction in the East Midlands

In conclusion the forms of transport dereliction in a regional context are examined, using the East Midlands as an illustration (Fig 29). Among other things this locality illustrates the effects which other forms of dereliction, notably mining subsidence, may have on the transport system. For example, the abandonment of the Chesterfield Canal above Worksop was caused by the closure of Norwood Tunnel after extensive damage caused by coal-mining subsidence, and similar damage to Butterley Tunnel caused the closure of the Cromford Canal. Subsidence has caused less spectacular damage to the railway network, but its effects can be seen on buildings and bridges situated along the lines which traverse the coalfields of the region. In addition canal and railway closures caused by losses of mineral traffic may be added to the sources of dereliction which have stemmed from the contraction of the mining industry. Much of the transport dereliction shown on Fig 29 is, however, a product of abandonment and closure which is not directly related to other

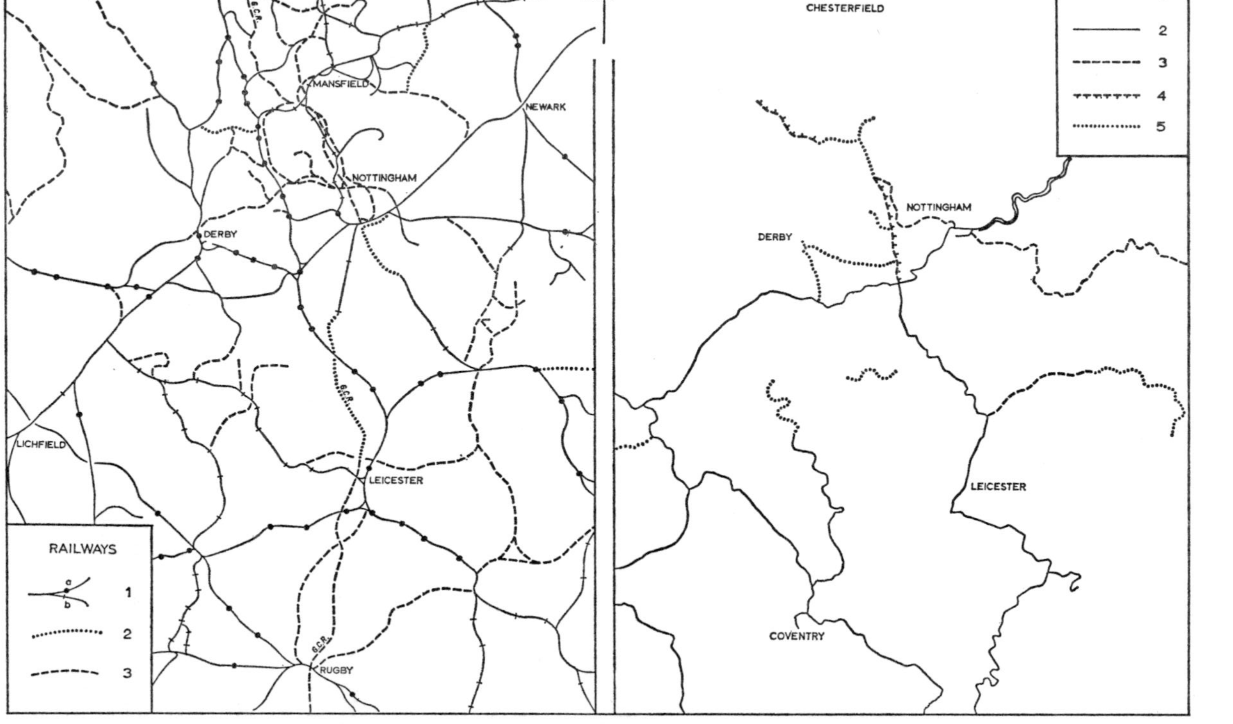

Fig 29 Elements of the derelict transport system in the East Midlands, 1970–72

Railways: 1, lines open to traffic; 1a, stations closed since 1948 on lines served by regular passenger services 1972; 1b, stations closed since 1948 on lines open to other traffic, 1972; 2, lines closed to traffic; 3, lines abandoned; GCR, Great Central Railway

Canals: 1, Trent Navigation (commercial waterway); 2, BWB cruising water ways; 3, BWB remainder waterways; 4, BWB remainder waterways likely to be restored as amenities; 5, abandoned canals. Data from various official sources

causes of dereliction but largely reflects changes in the volume and composition of canal and rail traffic over the years.

TABLE 10 Derelict land in Derbyshire, Leicestershire and Nottinghamshire resulting from canal and railway closures

County	*Canal and railway derelict land (ha)*	*Percentage of all derelict land 1966–7*
Derbyshire	164	12
Leicestershire	326*	58
Nottinghamshire	173	12
Total	663	20

Data for Derbyshire and Nottinghamshire from *Nottinghamshire and Derbyshire Sub-regional Study* (1968); for Leicestershire from the County Planning Officer

* Railway land only

Table 10 indicates the amount of transport dereliction in three East Midland counties during 1966–7. The high proportion of railway dereliction in Leicestershire, as a percentage both of the county's derelict land and the transport dereliction in the three counties, is particularly noteworthy. The closure before 1965 of extensive sections of rural track in east Leicestershire, where two-thirds of the railway dereliction was situated, was largely responsible for this state of affairs. Much of this derelict land had disappeared from the official returns (which is not to say it had been reclaimed) by 1970, when the amount of railway dereliction had fallen to 225ha: by that date the largest concentration was in west Leicestershire, notably along the route of the former Great Central Railway.

In each county abandonment has involved three elements in the railway network. Firstly, rural branch lines, exemplified by the once extensive system in the eastern parts of Nottinghamshire and Leicestershire (Fig 29). Secondly, branches in mining areas mainly but not exclusively built to carry coal traffic during the second half of the nineteenth century. Particularly striking examples are to be seen in the Erewash and Leen valleys, but there are also abandoned mineral lines in the

Leicestershire-South Derbyshire coalfield and on the ironstone field in east Leicestershire. The third category comprises sections of the main line system built in response to nineteenth-century competition among rival companies. The former Great Central Railway main line provides the outstanding example of this type. The section north of Nottingham via the Leen valley was abandoned during the 1960s and the track lifted. South of Nottingham passenger services survived longer, but a final decision to close the line was taken in 1969 and several sections of track were lifted during the early 1970s. In addition numerous stations were closed on routes which were kept open to traffic, adding to the number of potentially derelict sites.

The disused canals of the East Midlands also fall into three major groups. The first group comprises canals which have been drained and now exist as scrub-filled or marshy depressions forming linear parcels of derelict land. The long-disused Charnwood Forest canal, abandoned in 1799, provides one example, the Ashby Canal above Measham is another, and the greater part of the Derby Canal has fallen into this category since 1950. Secondly, there are canals which have been closed to navigation, parts of which may have been drained. This does not inevitably lead to dereliction: for example, part of the Oakham Canal is maintained for angling, but more commonly such canals are characterised by stagnant stretches of water, partial infilling with rubbish and a generally unkempt appearance. Much of the Nottingham Canal between Wollaton and Eastwood falls into this category. The third group comprises canals which, although closed to navigation, could be developed as amenity waterways. The Erewash Canal may be restored to cruiseway standards, and the Cromford Canal above the Butterley Tunnel is to be restored as an amenity. Unfortunately the section of canal through the tunnel and that running south to the junction with the Erewash Canal at Langley Mill cannot be restored, much of it having been drained and filled in during recent years.

This brief regional study indicates the considerable variety of transport dereliction and in particular emphasises its widespread occurrence through the linearity of the abandoned canals and railways. In many British counties derelict railway land forms a substantial proportion of the whole, and it might appear to be a particularly heavy burden on rural local authorities. This is not necessarily so, as the abandoned formation is often only a transient derelict feature, being amenable to simple and inexpensive forms of reclamation.

References to this chapter begin on p 315.

PART THREE

THE RECLAMATION OF DERELICT LAND

THE reclamation of derelict land has been mentioned at many points in previous chapters, for it is virtually impossible to speak of dereliction without an awareness of the problems of reclamation and the restoration of derelict land to beneficial uses. Strictly speaking reclamation means bringing land to productive use for the first time and restoration returning it to productive use after a lapse of time. In many reports on dereliction the terms are used synonymously and the same convention is followed in this account.

The introduction of an official definition of derelict land in Great Britain reflected the need to give guidance to those involved in financing its reclamation, and in practical terms the major purpose of recording the extent of dereliction is to assess the scale of the problem of reclamation which may exist. In this section reclamation is considered in terms of its areal extent and spatial distribution, its costs in relation to type of derelict land and proposed end uses after treatment, and the practical problems faced by reclamation agencies. The technical aspects of reclamation are, however, dealt with very briefly, not because they are unimportant but because this book is primarily concerned with dereliction as a facet of land use and cannot pretend to be a technical manual. Numerous technical papers exist on the reclamation of derelict land, although, as the county planning officer for Lancashire has recently observed, 'the widespread dissemination of the basic technical problems [of restoration] and its longer term results has never been successfully achieved' (Rowbotham 1973). Two important sources of technical data deserve specific mention. J. R. Oxenham's *Reclaiming Derelict Land* (1966) is a compendious account drawing on more than 20 years' experience of derelict land reclamation. The more recent two-volume report entitled *Landscape Reclamation* (1971–2), produced by a team of research workers at the University of Newcastle upon Tyne, covers many topics but is particularly concerned with the botanical and pedological aspects of reclamation projects, and the design and management of reclaimed sites. Two international symposia on spoiled and degraded land in Europe held in 1973 dealt with technical aspects of reclamation among other things, but their proceedings were published after this volume had been written. There are in addition several bibliographies of technical papers which are listed as additional references to this section on p 320.

CHAPTER SEVEN

Reclamation in Great Britain: General Review

RECLAMATION BEFORE 1945

BEFORE 1945 there was no concerted effort to reclaim derelict land in Great Britain. Such reclamation as did occur was piecemeal and much of it may not have been considered as such when it took place. The government did not normally finance schemes of reclamation before 1945, and was sparing in its allocation of funds even until the mid-1960s. Thus in the 1930s, when much of the hard core of dereliction came into being and there was an opportunity to provide relief work for the unemployed by introducing a programme of reclamation, there was governmental antipathy rather than interest. On several occasions during the 1930s attempts were made to secure parliamentary sanction for state-aided reclamation schemes, but none were successful. The annual reports of the commissioners for the Special Areas reveal a similar attitude, for it was established that derelict land would not be cleared unless there was a prospect of attracting new industry by doing so. In 1936 the commissioner for England and Wales stated that he was 'not prepared to sponsor schemes [of reclamation] whose sole object is to provide employment'. Consequently the reclamation of derelict land was confined to a small number of industrial sites, notably on the banks of the Tyne between Gateshead and South Shields and in Merthyr Tydfil (see p 256). By 1938 the commissioners had begun to relent a little by encouraging modest schemes of tree planting on colliery spoil tips in Durham and central Scotland, but it is quite clear that no major state-financed programme of reclamation was ever envisaged. Although the 1944 Town and Country Planning Act gave local authorities specific powers to acquire and reclaim derelict land, the use of these apparently broad provisions was discouraged by an administrative stratagem. Land which had a value was considered not to be derelict, so that no grant was available for its treatment. On the other hand, if the land had no value, there could be no financial loss in putting it to a more beneficial use, and no grant was payable in this case either. Not surprisingly the provisions of the Act were scarcely used and they were repealed, never to be replaced, in 1947 (Oxenham 1969).

Consequently almost all the reclamation of derelict land carried out before 1945 was either privately financed or funded by local authorities and fell into one of three categories. Firstly, there was the use of derelict land for building purposes, which was probably very rarely recognised as formal reclamation. The extension of factory and mine buildings on to waste land and the construction of houses on the sites of shallow clay pits were common forms of this type of reclamation from the mid-nineteenth century onwards. Secondly, there was reclamation of derelict land for recreational purposes and, thirdly, there was tree-planting on the spoil heaps of various types of mine: numerous examples of both forms of reclamation exist for the period before 1945 and two instances drawn from the Black Country will serve as illustrations. This region contained an excessive amount of derelict land by the 1870s, a period of severe unemployment in the area (see p 283). Some local authorities sought a solution to both problems by using the unemployed to construct public parks on derelict land, at least five being opened between 1884 and 1900. For example, East Park, Wolverhampton, opened in 1896, was located on the site of Chillington ironworks and an adjacent tract of low spoil heaps and subsidence hollows, covering 20ha in all. Most of the land was donated, so that the expense of reclamation (£14,000) was largely accounted for by the costs of levelling the site and laying it out as a park. For many years East Park was a curious land-use anomaly, an oasis in a desert of dereliction; even in 1971 its south west perimeter was still bounded by derelict land.

The same public interest which supported the creation of parks was also influential in backing the activities of the Midland Reafforesting Association, founded in 1903 (Wise 1962). The association considered that 5,670ha of derelict land were then suitable for afforestation and that ultimately 12,145ha might be planted. In fact only 33ha of afforested derelict land survived when the association was dissolved in 1924, almost half of it in a single plantation at Moorcroft Hospital, Moxley. Some plantations had failed to take root owing to atmospheric pollution and unsuitable soils, and vandalism was also responsible for several failures, but it is no slur upon the aspirations of the association to suggest that a well-meaning voluntary body was incapable of meeting the ambitious target which it had set itself.

THE STATE AND RECLAMATION, 1945–73

Although private schemes of reclamation had given some indication of both the possibilities for and the drawbacks of treating derelict land, any large-scale attack on the problem of dereliction clearly needed the legislative and financial support of the state. Various enactments between 1945 and 1966 included provisions which might be used in the reclamation of derelict land, notably in Acts relating to the distribution of industry, the creation of national parks and the financing of local authority housing (Oxenham 1966, Clark 1969). In fact very little derelict land was reclaimed under the terms of such legislation, with the exception of that used for housing in the Black Country. For example, only 40ha of derelict land were reclaimed in Wales between 1960 and 1966, whereas under the provisions of new legislation 153ha were restored in 1967 alone. The greatest stimulus to schemes of reclamation was provided by the Industrial Development Act (1966), the Local Government Act (1966) and the Local Employment Act (1970), for these combined to enhance the rate of grant available for reclamation and, more important, to increase the territorial coverage of state aid. In addition many other Acts on the statute book are used to enable local authorities to acquire derelict sites in order to qualify for state aid. Land in private ownership does not normally qualify for reclamation grants, except under the provisions of the Special Environmental Assistance Scheme (SEAS) introduced in the assisted areas for one year in June 1972. On the other hand derelict land in private ownership is still reclaimed at no cost to the Exchequer, mainly for the construction of factories and housing, although information on such projects is less easily come by than for those which are publicly financed.

The level of grant available under current legislation varies according to locality. In Development Areas the rate of grant fixed in 1966 was 85 per cent of the approved net costs of reclamation, and this can be made up to a maximum of 95 per cent by rate support grants. From 1970 onwards, 75 per cent grants were made available in the Intermediate Areas and the same level of grant is payable in National Parks and Areas of Outstanding Natural Beauty and in the Derelict Land Clearance Areas (Fig 31). Elsewhere the rate of grant is 50 per cent of approved net costs, which include those of acquiring the land, preparing plans for its treatment, and all rehabilitation work. Any revenue secured from materials salvaged during reclamation or from subsequent use of the land is deducted from the grant, but not until the reclaimed site attracts income. In addition to the grants administered directly by the state for purposes of reclaiming derelict land local authorities are able to secure other funds and county councils are able to give assistance to their poorer areas out of general rate income. The principal point to emphasise is that since 1966 substantial grants have been widely available for the reclamation of derelict land, and that these were not subject to the restrictions placed on public spending in the early 1970s. This reflects the importance which the state attaches to its declared aim of reclaiming all derelict land during the 1970s.

Although the state has borne much of the financial burden of reclaiming derelict land, the responsibility for framing and implementing schemes of reclamation has always rested with local authorities and will continue to do so after the reorganisation of local government in 1974. It was suggested in 1960 that the Opencast Executive of the NCB might also function as the central derelict land reclamation agency, and this idea was given fresh currency in 1970 (Civic Trust 1970), only to be rejected by Parliament in the following year. A compromise has been achieved whereby the technical expertise of the Opencast Executive is used to plan and direct many of the major schemes of reclamation, but within a programme laid down by the local authorities. As there is also central control over the funding of reclamation projects, it would seem to be pointless to create a central derelict land reclamation agency, which would then have to coordinate its programme of reclamation with other planning decisions taken by local authorities.

TECHNICAL DEVELOPMENTS IN RECLAMATION

Most of the limited amount of reclamation carried out before 1945 relied on manual labour to level spoil tips and infill pits, and this limited the scope of reclamation projects. The development of heavy mechanical equipment for civil engineering and opencast mining provided machinery that was equally well suited to the reclamation of derelict land. J. R. Oxenham (1966) has outlined the principal types of equipment available, most of it introduced to Great Britain during the period 1938–45. These machines in combination have revolutionised the tasks of shifting and regrading spoil heaps, filling pits and hollows, and transporting materials within and between the sites of reclamation projects. A second major area of technical development concerns the rehabilitation of derelict land after the major civil engineering works have been completed. The importance of adequate on-site drainage has been recognised as crucial to successful grassland management on reclaimed derelict land, whether for agricultural or recreational purposes. Much has also been done to identify the processes of soil formation on various types of waste material, thus enabling reclamation to proceed without the need to import top soil. The planting of trees in raw shale received particular attention when it was necessary to devise inexpensive forms of reclamation before the more widespread availability of generous grants in the mid-1960s, and much valuable pioneer work was carried out in Lancashire (Casson and King 1960) and in County Durham. Concurrent research into techniques of directly seeding waste materials such as colliery shale and PFA had a twofold importance: firstly, it became possible to establish grass directly on regraded colliery spoil heaps and PFA dumps. Secondly, these materials could be used as fill at other reclaimed sites without the necessity of employing more costly soil-forming materials. Before the mid-1960s it was common practice to cover spoil heaps and PFA dumps with a thin layer of top soil (between 15–25cm) before seeding, whereas direct seeding is now commonplace. The use of such small amounts of top soil may not seem to have been a very great impediment to reclamation, but it must be remembered that it was exceedingly expensive in relation to total costs of reclamation. In a pioneer report on the establishment of grass on raw shale (Coates 1964) it was noted that the cost of covering 1 hectare with 23cm of topsoil was approximately £300, even if soil was available on site, whereas the cost of preparing and seeding raw shale was only £90 per hectare.

A third major technical development has stemmed from a

growing awareness of the need for careful management of reclaimed land. The early schemes of afforestation in the Black Country were largely negated by bad management of the plantations, and recent experience in urban areas has shown that schemes must be designed to minimise the chances of vandalism and depredation by domestic animals. Advances have also been made in the management of soils and grassland after reclamation in the attempt to ensure that the long-term establishment of reclaimed land is not damaged by short-term errors stemming from lack of maintenance. It must be stressed that all three facets of improved reclamation technique are equally important. The introduction of heavy machinery capable of changing the derelict landscape overnight may have seemed the most spectacular development, but without the parallel discovery of cheap means of establishing vegetation and the establishment of proper management practices, much of the impressive civil engineering work would have been abortive. Reclaimed sites may take as long as 20 years to mature, and unfortunately dramatic capital works programmes often produce landscapes which are bleak and disappointing to the layman, who had been led to expect something much better. Consequently it is sometimes difficult to persuade the public that a particular site has been reclaimed; there may be a recurrence of long-established illicit waste dumping, and grass, trees and shrubs may be trampled and destroyed. In fact the site may be treated as casually as its derelict predecessor. There is, therefore, much more to successful long-term reclamation of derelict land than unleashing mighty machines in order to devour spoil heaps and to fill abandoned quarries.

Finally, it should be noted that technical advances have also been made in minimising the amount of newly created derelict land. The immediate reclamation of various types of surface mineral working provides a case in point, not only reducing the amount of new dereliction but also advancing the development of reclamation techniques generally. The introduction of mining and quarrying methods which would prevent the formation of derelict land has been less widespread, particularly in the deep mining of coal. However, some progress has been made, notably in tipping colliery waste in a manner which permits reclamation of parts of the site while the remainder is in active use. In addition the drafting of planning consents for new mineral workings has become more exact in specifying the kind of rehabilitation which must follow abandonment, and this in itself represents a substantial advance in the reduction of the volume of new dereliction.

THE PROGRESS OF RECLAMATION

The statistics relating to the reclamation of derelict land are collected on the same basis as those covering dereliction as a whole. The official returns of reclamation relate only to schemes carried out on land that was officially defined as being derelict. Hence, reclamation of mineral-working sites immediately after cessation of operations is not included in the returns, which thereby exclude a substantial amount of work carried out at opencast coal and ironstone mines, and at sand and gravel pits. More important the returns are occasionally ambiguous in their definition of what constitutes reclamation. If the returns for successive years show a net decrease in derelict land, this cannot automatically be taken as an indication of reclamation, for some counties have undertaken reappraisals of derelict land which have produced apparently dramatic reductions in its extent at the stroke of a pen. However, the returns also include separate statistics on the amount of land reclaimed in each year, together with a statement of projected and potential reclamation. As far as can be determined, the returns of land reclaimed are accurate, although one local authority returned 115ha of land reclaimed when it would have been more accurately described as reappraised.

The returns of projected reclamation indicate the amount of land which a local authority intends to reclaim during the year following the survey, but these data are particularly suspect. In

England they represented an overestimate of 29 per cent of the reclamation completed during the period 1968–71, much of which stemmed from legal and technical problems encountered in carrying out reclamation projects to a strict schedule, for there is much that can delay progress and very little that can accelerate it. The returns also identify the proportion of derelict land which justifies restoration. It is not clear whether all local authorities use the same criteria in determining which types of derelict land justify treatment, but there is almost certainly a difference of interpretation in relation to derelict railway land in rural areas, much of which is put to beneficial use without the intervention of a formal scheme of reclamation; indeed, some local authorities exclude it from the category of land justifying restoration for this reason, while others include it. Many of the anomalies relating to reclamation will be resolved by the new form of survey to be introduced in 1974, but the data currently available need to be interpreted with caution.

Reclamation in Great Britain

The process of reclamation in Great Britain since 1964 is summarised in Table 11. During that period 25 per cent of the derelict land recorded in England and Wales in 1964 was reclaimed, as was about 13 per cent of that in Scotland. The full extent of dereliction in Scotland has never been accurately surveyed and the progress of reclamation has been officially described as 'bitterly disappointing'.

The distribution of reclamation in England between 1968–71 is shown in Fig 30a and summarised in Table 12. During this period the annual rate of reclamation trebled, largely owing to the impetus provided by legislation passed in 1966. Not surprisingly the pattern of reclamation closely resembled that of dereliction (Fig 13), with the important exception that Cornwall undertook very little reclamation during this period.

There is a general correspondence in each region between share of the national total of reclamation and of the national total of derelict land, though there are some exceptions to this

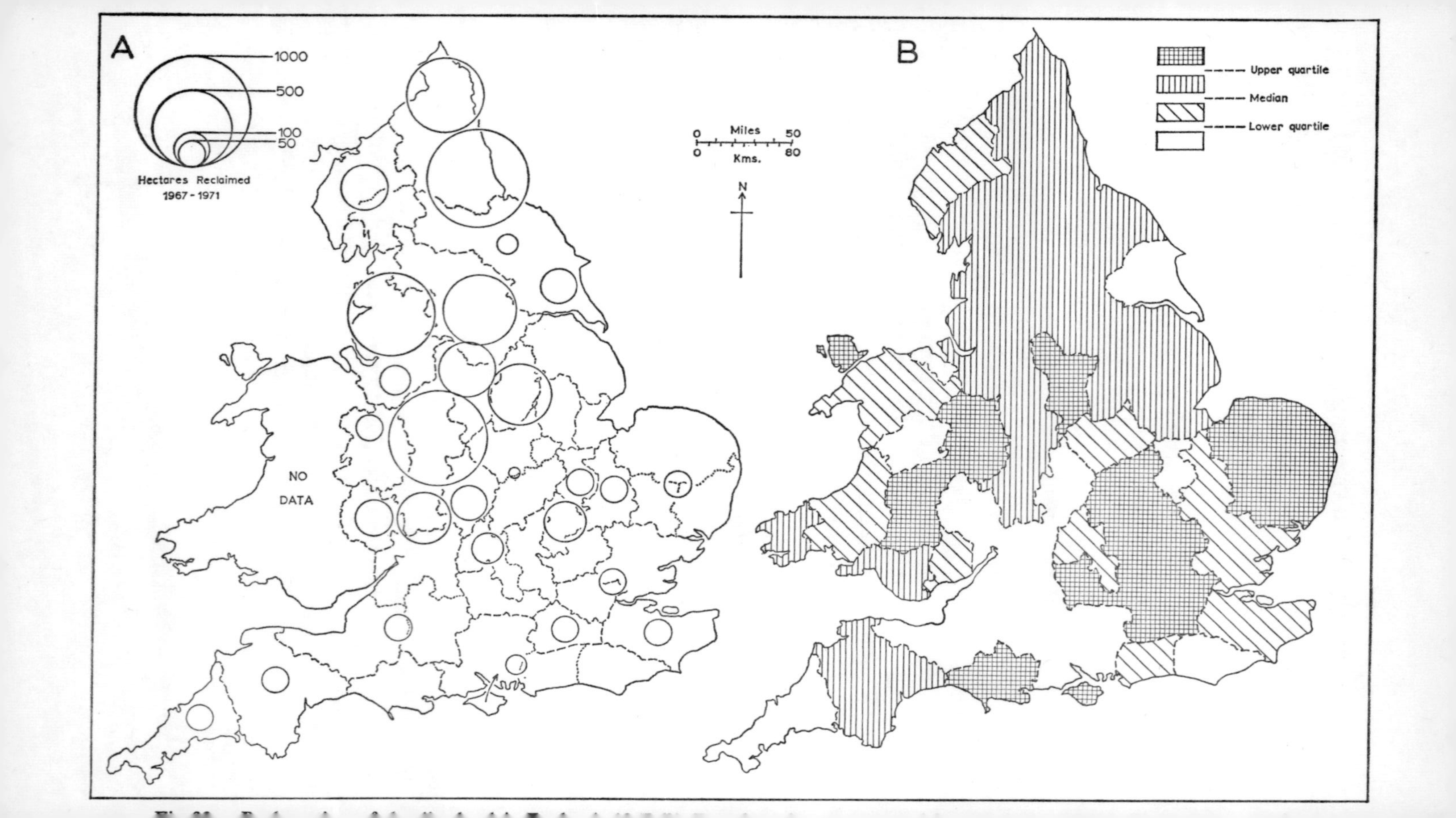
A
1000
500
100
50
Hectares Reclaimed
1967 - 1971
NO
DATA
0
Miles
50
0
Kms.
80
N
B
Upper quartile
Median
Lower quartile

TABLE 11 Reclamation of derelict land, Great Britain, 1964–71

	Hectares	
	England and Wales	*Scotland**
1964	992	32
1965	957	94
1966	737	114
1967	747	66
1968	1,078	61
1969	1,321	120
1970	1,786	440† (1970–71)
1971	2,332	
TOTAL	9,950	927

Data from Department of the Environment, Welsh Office, Scottish Development Department and *Hansard*

* Grant aided schemes only

† To 30.6.72

TABLE 12 Regional distribution of reclamation of derelict land, 1968–71

		Land reclaimed, 1968–71				*Proportion of national total*	
Region	(a)	(b)	(c)	(d)	(e)	(f)	(g)
North West	620	12·0	40 (34·5)	16 (25·0)	44 (40·5)	12	14
North	1,456	16·5	45 (30·0)	8 (18·0)	47 (52·0)	28	22
Yorkshire and Humberside	690	15·5	33 (24·0)	29 (42·0)	38 (34·0)	13	12
East Midlands	536	16·5	48 (27·0)	21 (31·5)	31 (41·5)	10	8
West Midlands	1,262	23·0	29 (38·0)	19 (16·0)	52 (46·0)	24	14
South West	158	2·0	16 (75·0)	28 (10·0)	56 (15·0)	3	21
East Anglia	147	12·0		55 (36·0)	45 (64·0)	3	3
South East	414	19·0	0 (3·0)	83 (69·0)	17 (28·0)	7	6
England	5,283	13·5	34 (37·0)	23 (26·0)	43 (37·0)	100	100

(a) area reclaimed (ha), (b) percentage of derelict land in 1968 reclaimed by 1971; percentage of derelict land reclaimed accounted for by (c) spoil heaps, (d) excavations and pits, (e) other dereliction (figure in brackets percentage of all derelict land in each category), (f) reclaimed land, 1968–71, (g) derelict land, 1971

Data from Department of the Environment; see also Table 13 for reclamation in 1972

general conformity. The North region had a markedly greater share of reclaimed land, largely because the whole area enjoyed Development Area status, with the highest rate of reclamation grants, and because of a particularly vigorous campaign of reclamation in County Durham. Even greater disparity is to be seen in the West Midlands, though only North Staffordshire enjoyed the special status of a Derelict Land Clearance Area (Fig 31). In the West Midlands conurbation the high rate of reclamation of derelict land reflected market demand for land rather than the influence of state-financed programmes of reclamation. Conversely the South West region had reclaimed very little land in relation to its share of the national total of dereliction, in spite of the fact that parts of the region enjoyed Development Area status or qualified for enhanced grants on other grounds. The impact of state financial aid on the amount of reclamation carried out is clearly variable, being an obvious stimulus in much of northern England but of little effect in the south west region. Conversely many parts of the midlands and south east England were able to reclaim a substantial amount of their derelict land because normal market opportunities offset the disadvantage of having the lowest rate of state financial aid.

TABLE 13a Reclamation carried out during 1972; sources of finance by type of dereliction

	Amount reclaimed (ha)	*Percentage of derelict land existing in 1971*	*Percentage of land reclaimed with state finance*	*by local authorities**	*By other bodies*
Spoil heaps	866	6·0	80	5	15
Excavations and pits	200	2·0	45	30	25
Abandoned British Rail land	232	7·0	63	16	21
Other dereliction	872		77	6	17
TOTAL	2,170	5·5	74	9	17

* Excluding grant-aided schemes

Data from Department of the Environment

TABLE 13b Reclamation carried out during 1972; sources of finance by regions

	Amount reclaimed (ha)	*Percentage of national total*	*Percentage of derelict land existing in 1971 reclaimed*	*Percentage of land reclaimed with state finance*	*by local authorities**	*By other bodies*
North West	405	18	6·5	82	12	6
Northern	724	33	9·0	91	2	7
Yorkshire and Humberside	301	14	5·0	61	12	27
East Midlands	285	13	8·0	79	5	16
West Midlands	342	16	7·0	50	13	37
South West	36	} 6	0·5	53	20	27
East Anglia	34		2·5	14	17	69
South East	43		2·0	0	67	33
TOTAL	2,170	100	5·5	74	9	17

* Excluding specific grant-aided schemes

Data from Department of the Environment

More precise data on the amounts of derelict land reclaimed by various agencies exist for 1972, when the first stage of the new survey of derelict and despoiled land (see p 35) collected information on the funding of restoration schemes (Table 13). In 1972, 2,170ha of derelict land were reclaimed, equivalent to 5·5 per cent of that recorded on 31 December 1971 (there are no returns of derelict land existing at the end of 1972 and the net increase or decrease during the year cannot therefore be calculated). Almost three-quarters of the land reclaimed was covered by specific grants in aid from the Department of the Environment, and a further 9 per cent of the area was restored by local authorities who may have secured other forms of state finance. The remaining 17 per cent (375ha) was restored by other forms of funding, including that incidental to private development.

The regional pattern of reclamation in 1972 differed little from that of preceding years, although the proportion of the national total of restored land situated in the Northern and

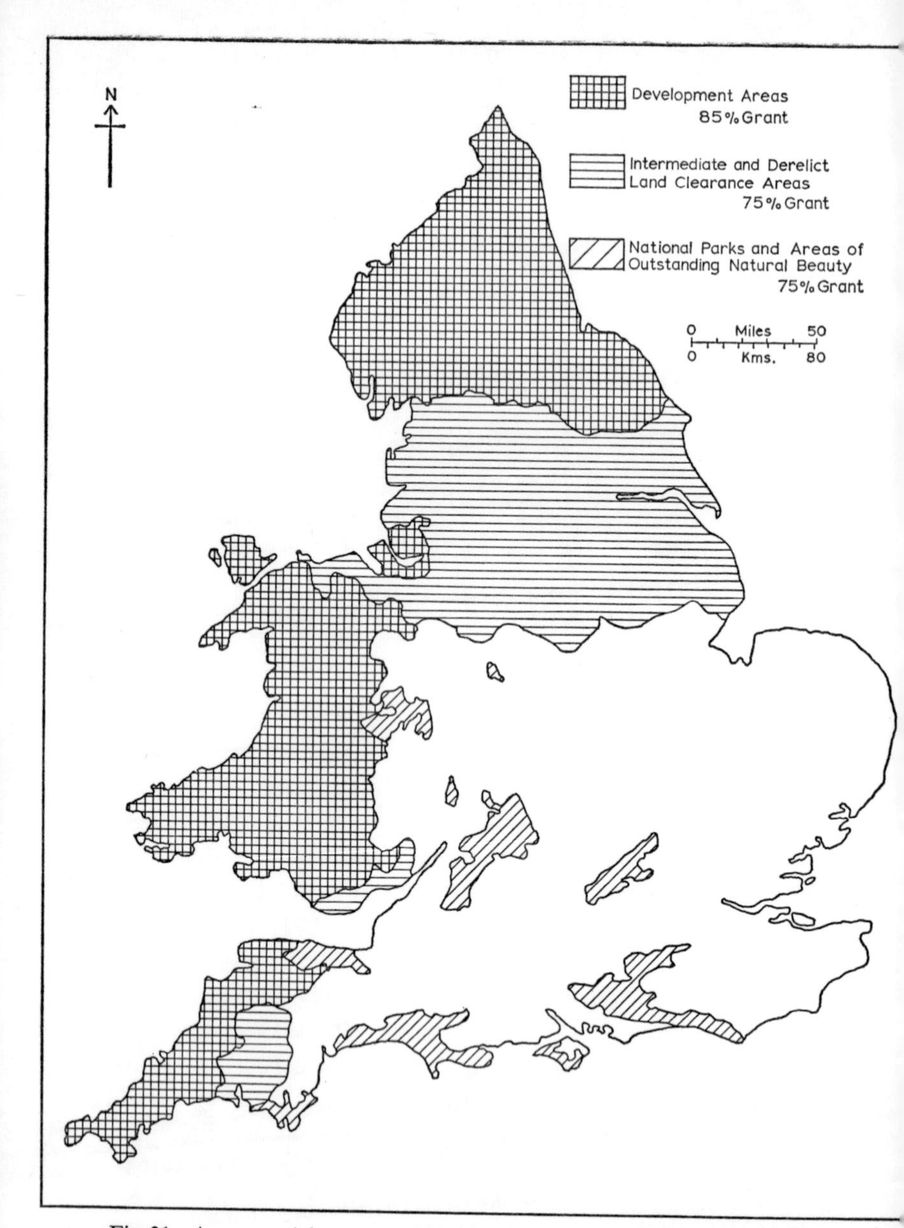

Fig 31 Areas receiving grants for the reclamation of derelict land, 1972. In addition the whole of Scotland qualifies for the 85 per cent rate of grant (except the Edinburgh area, 75 per cent). Localities not separately identified receive a 50 per cent grant

North Western regions (52 per cent) was substantially greater than in 1968–71 (40 per cent). A very high proportion of the land reclaimed in these two regions (88 per cent) was funded by specific grants in aid, compared with 7 per cent funded by the private sector. Conversely a very low proportion (6 per cent) of the reclaimed land was situated in southern England (East Anglia and the South West and South East regions), of which 21 per cent was funded by specific grants in aid and 42 per cent by the private sector. Put another way, the north contained 55 per cent of the land reclaimed with specific state aid but 36 per cent of the derelict land in England at the end of 1971, while the south contained 1·5 per cent and 30 per cent respectively. In the remainder of England 62 per cent of the derelict land was restored with specific grants in aid and 27 per cent was privately funded. Although the returns for this area do not adequately distinguish between reclamation schemes located in the assisted areas (Development, Intermediate and Derelict Land Clearance areas – Fig 31) and those situated elsewhere, the evidence strongly suggests that the assisted areas attracted a high proportion of state aid and contained an equally high proportion of the reclamation projects. The 1972 returns generally confirm the arguments put forward in explaining the pattern of reclamation during the period 1968–71, notably in revealing the significance of specific state aid in the assisted areas (except those of the South West region) and of other forms of expenditure elsewhere in England.

The identification of the proportion of derelict land in each category justifying restoration provides data on the potential for reclamation in each local authority area, assuming that the returns genuinely reflect the situation. In 1969–70, 66 per cent of the derelict land in England and Wales justified restoration: 63 per cent of spoil heaps, 51 per cent of pits and excavations, and 76 per cent of other forms of dereliction. There is no well defined relation between the amount and density of dereliction and its propensity for reclamation, nor does the threefold structure of derelict land in a given area relate clearly to the potential for reclamation (Fig 30b). Thus Cornwall, with the highest amount and density of derelict land, also had one of the lowest rates of potential for reclamation (11 per cent of derelict land justifying restoration), whereas Derbyshire, another county with a high density of derelict land, had a high potential for reclamation (94 per cent). Conversely Dorset and Berkshire, counties with very low densities of dereliction, had very high rates of potentially reclaimable land (95 per cent in both counties), but Gloucestershire and Wiltshire, with similarly low densities of derelict land, recorded low rates of potential for reclamation (33 per cent and 36 per cent respectively).

The pattern of potential for reclamation cannot be simply explained, but three general points may be made about it. Firstly, several counties with severe dereliction assessed the potential for reclamation to be 70 per cent or more of their derelict land, with many of the counties in northern and midland England falling into this category. Secondly, the potential for reclamation was often surprisingly low in southern England, even in some counties bordering on Greater London, where pressure on space for all forms of development is considerable. Closer examination of the returns suggests that several counties in this area interpreted the term 'justifying restoration' very narrowly. Thus Oxfordshire was pessimistic about the likelihood of its gravel pits being reclaimed, even though, on the basis of past experience, many of them would be adopted for amenity use. Both Hampshire (3 per cent) and Herefordshire (2 per cent) returned exceedingly low estimates of the potential for reclamation, but in both counties this reflected the fact that railway land formed the bulk of the dereliction. Indeed Herefordshire 'reclaimed' 60 per cent of its derelict railway land in 1970, notwithstanding this pessimistic assessment of potential for reclamation, by regarding the sale of track to adjacent landowners as a form of restoration. Thirdly, a separate analysis of the returns from English county boroughs shows that their assessment of the potential for reclamation of derelict land (93 per cent of all recorded dereliction) is predictably much higher than that of the administrative counties (62 per cent), since derelict land is much more likely to be reclaimed in the major urban areas than elsewhere. This is particularly true of midland and southern England, where 98 per cent of the derelict land in county boroughs (including the entire Greater London Council area) was considered worthy of restoration.

Costs of reclamation

However imperfect the statistics relating to reclamation may be, they are at least superior to those on costs, which do not exist in a standard form for the entire country. The data used in this section have been assembled from a variety of sources and authorities, not all of which present their financial returns in the same way. The most commonly quoted average costs of reclamation are those based on state-aided schemes, which form the bulk of the total volume of work carried out. Before 1966 the limited amount of work carried out under the provisions of the 1947 Town and Country Planning Act and similar legislation cost an average £1,600 per hectare. In 1968–9 the gross costs of grant-aided reclamation schemes were £4,940 per hectare in Wales, £3,200 per hectare in England and £2,850 per hectare in Scotland. The differences in cost between schemes are the product of several variables, including costs of acquiring derelict land, of civil engineering works, and of managing the sites immediately after reclamation. In turn these variables depend on the location of the site, which affects its market value, its physical characteristics and the use to which it is put after reclamation.

A particularly comprehensive survey of projected costs of reclamation exists for Cumberland, based on the county's programme of reclamation for 1971–80: all cost estimates are at 1971 prices, so that no allowance has to be made for inflation, as is the case with returns relating to reclamation carried out over a period of time in the past. The area to be reclaimed amounts to 860ha, at an estimated gross cost of £2,740 per hectare. In almost every category of reclamation scheme the median and the modal cost of reclamation lies between £3,000 and £3,250, irrespective of the type of dereliction and the proposed use of the site. There are, however, major variations within most of the categories, which are indicated in Table 14. The costs range from the remarkably low £346 per hectare estimated for the conversion to a marina of the flooded Hodbarrow iron mines, to the expensive conversion of the Ropewalk, Cockermouth, into a public open space (£16,432 for a half hectare site). That both should be amenity schemes is in itself adequate warning against generalising about the costs of reclamation. Indeed the main lesson of the Cumberland survey is that although both the modal and the median costs may closely approach the national average, the variation about these values is not based on any simple relation between the type of derelict land and the end use envisaged after reclamation.

Similar conclusions may be drawn from other less compre-

TABLE 14 Range of costs of reclaiming derelict land, by major categories, Cumberland (1971–80 programme)

	Type of dereliction					*Average cost per hectare*
Restored to	*Colliery*	*Other mining*	*Industry*	*Quarrying*	*Other*	(£)
Agriculture (63)	593–9,884 (24)	741–6,177 (16)	3,089–4,695 (2)	371–9,143 (18)	988–3,089 (3)	2,542
Amenity (44)	1,235–6,177 (9)	2,471–3,286 (4)	346–11,737 (8)	679–4,942 (8)	741–32,864 (15)	4,387
Forestry (44)	593–3,706 (6)	593–3,212 (16)	1,235 (1)	494–4,940 (17)	494–1,235 (4)	2,227
Industry (12)	1,606–3,459 (3)	3,088–4,942 (2)	2,718–12,355 (3)	1,235 (1)	1,977–3,089 (3)	3,820
Other and mixed uses (22)	4,942–11,120 (4)	1,235–3,706 (5)	1,112–9,884 (2)	618–1,235 (4)	1,235–6,795 (7)	1,657
All uses (185)	593–11,120 (46)	593–6,177 (43)	346–12,355 (16)	371–9,143 (48)	494–32,864 (32)	2,740

The range of figures indicates the highest and lowest cost in each category: the figure in brackets indicates the number of reclamation projects

hensive surveys and reports produced by various local authorities. The high cost of reclamation in urban areas is confirmed by an investigation of the detailed confidential schedules used in the compilation of Table 15. The costs cited below show that in a major English city costs of acquisition accounted for 28 per cent of the total cost of reclamation in spite of the fact that almost

TABLE 15a Costs of reclaiming derelict land, County Borough X, 1970–75

	£	
(a) *Initial basic costs*		
Land acquisition	709,230	
Preparatory engineering	713,950	
Treatment and landscaping	1,040,360	
Salaries/overheads	73,590	
TOTAL	£2,537,130	(£15,230 per ha)
(b) *Subsequent costs*		
Loan charges	£351,400	
Site maintenance	£55,640	

half of the derelict land was in corporate ownership *before* the reclamation programme was mounted. The acquisition of the balance of the derelict land cost £6,470 per hectare, compared with an average reclamation cost of £15,230 per hectare. It may be wondered why this city would wish to embark on such expenditure, particularly as the bulk of the land in this case was destined to be used as open spaces. One reason is that state aid will provide at least 50 per cent of the cost, and in County Borough X it provided 75 per cent, as it lay within a Derelict Land Clearance Area. Another explanation is that the reclamation schemes all lie in parts of the city which are chronically short of recreational space, and the programme is well merited for social reasons. But, perhaps most important of all, the costs of reclaiming derelict land are roughly equal to those of acquiring equivalent areas of other land in the same localities without the added expense of converting the sites to new uses. Consequently, reclamation of derelict land is an attractive alternative to other courses of action in densely urbanised areas, in spite of the apparently high costs.

Data from schedules prepared by several county boroughs in Lancashire show that where costs of acquisition are separately identifiable, they occasionally exceed the costs of subsequent reclamation (Table 15b). Schedule 2 demonstrates the fallacy of assuming that waste heaps and abandoned excavations in close

TABLE 15b Schedules of reclamation projects from selected Lancashire county boroughs

Scheme	*Extent (ha)*	*Cost of acquisition*	*Gross cost of reclamation**	*Total gross cost per hectare*
		(£)	(£)	(£)
1 Railway land	3·3	16,000	12,000	88,485
2 Chemical waste	11·0		239,000	21,727
Clay hole	6·0	135,000	56,000	31,833
3 Chemical waste and railway land	5·9		35,000	5,932
4 Bleach works	7·0		100,000	14,285
5 Tannery sludge	1·3		9,000	6,923
6 Abandoned canal	4·3	37,700	71,250	25,337
7 Cemetery	1·0	+	21,000	21,000
8 Brickworks and claypit	16·0	150,000	156,000	19,125
9 Allotment gardens	0·5	+	720	1,440
10 Railway land	5·0		175,000	35,000

* Includes cost of acquisition unless otherwise stated; +land already in municipal ownership

Data from returns supplied by various County Borough Planning Offices

proximity provide a simple solution to problems of reclamation. The chemical waste was to be levelled and dumped into the clay pit at a cost of £430,000 in order to provide 17ha of reclaimed land. The high cost of reclaiming railway land in urban areas is also indicated, in sharp contrast to the common practice in rural areas of writing it off without public expenditure. For example, schedule 10 represents the cost of converting part of the abandoned railway system of a large city into public footpaths.

Uses of reclaimed derelict land

Data on the end uses to which reclaimed derelict land has been put are not collected systematically, but some information is presented in Table 16, relating to post-war reclamation schemes in a group of urban localities. Housing accounted for the highest proportion, largely because of the programme of local-authority house construction on derelict land in the Black Country. Industrial uses were next in importance, both in economically prosperous areas and in those localities where derelict sites were reclaimed to provide state-financed industrial estates. Open-space and recreational land accounted for one-fifth of the total, as many urban areas used the opportunity provided by reclamation schemes to add to their stock of

TABLE 16 End uses of reclaimed land in urban areas, England, 1946–70

	Hectares	*Percentage*
Housing	1,806	43
Industry	953	23
Open space and recreation	824	20
Other uses	369	9
Agriculture	235	5
TOTAL	4,187	100

Data from various planning reports

amenities, particularly in poorly endowed nineteenth-century industrial localities. The category 'other uses' included land upon which schools and other institutions were built, and it is probable that some of the land originally restored to agriculture was subsequently converted to recreational and institutional uses. Current programmes of reclamation appear to put a much higher premium on restoration for amenity and agricultural uses, but this may in part reflect uncertainty about the ultimate function of some sites. Clearly, however, an increasing proportion of the derelict land to be reclaimed will answer a variety of recreational and general amenity purposes, even though this may produce a low financial return in comparison with use for either industrial development or housing. This partly reflects the stimulus of a generous policy of granting state aid without requiring a guarantee of financial returns, but it is also a reflection of the sustained demand for various types of recreational space.

References to this chapter begin on p 315

CHAPTER EIGHT

The Reclamation of Derelict Land: Case Studies

MANY of the general points made in the previous chapter can best be illustrated by specific case studies, several of which also relate to the study of dereliction forming Part Two of this book. The method of approach adopted is to take the three categories of dereliction identified in Part Two as the basis for selected studies of reclamation and to conclude with a regional study of recent progress in the reclamation of derelict land in South Wales. Further examples also appear in Chapter 9 as part of a more comprehensive study of dereliction and reclamation in North West England and the West Midlands.

RECLAMATION OF MINING DERELICTION

As the principal source of mining dereliction, coal mining has long attracted schemes of reclamation. When the derelict land created by the industry was of low amplitude the physical problems of reclamation were slight, but the incentives to reclaim were also negligible. Tree planting on spoil heaps was the most widely adopted form of reclamation, but even this simple form of restoration was not common. Even though areas of low-amplitude dereliction present few technical problems for modern methods of reclamation, substantial pockets still survive along the north crop of the South Wales coalfield and in eastern Shropshire. In the latter area much of the derelict land was deliberately included within the designated area of Telford New Town so as to make use of reclaimed sites instead of encroaching on agricultural land. The consultants' report of 1966 indicated that there were 3,035ha of land affected by shallow mining in the locality (Madin 1966), but by 1972 very little of it had been built over and the major concentration of derelict land north of the Holyhead Road remained largely intact (Tolley 1972). One of the most serious obstacles to using the long-abandoned sites of collieries for building is the uncertainty about the location of shafts and the methods used to seal them on abandonment, which add unsuspected hazards to what would otherwise be very simple reclamation projects.

The majority of the reclamation schemes in recently abandoned coal-mining districts have three elements of dereliction to contend with: abandoned pit-head installations, spoil heaps and slurry ponds, and land damaged by subsidence. Abandoned plant and buildings present few difficulties that are not also encountered in reclaiming heavy industrial sites, except that there is a need to seal unused shafts. Spoil heaps of the ridge or cone variety present the greatest technical problems, since it is rarely possible to reclaim them without major engineering works, although their removal may be aided by working the spoil for residual coal or as a source of shale. The principal technical difficulty at many sites has been spontaneous combustion of apparently dormant tips once air has been admitted to the compacted spoil. Care has also to be taken that the levelled spoil is properly sealed after treatment, or there is the danger that spontaneous combustion may recur beneath the reclaimed sites. Slurry ponds at washeries are sometimes worked for coal; but if not, their main problem is that the unconsolidated silt has poor load-bearing properties. Land damaged by subsidence has frequently been reclaimed by tipping colliery waste into the subsidence hollows and, in at least two instances (at Rugeley and Knottingley), PFA from adjacent power stations has been tipped on to subsiding land as part of a planned programme of reclamation. There are also reclamation schemes which leave the flooded areas intact, eg the wildlife sanctuary at Polesworth, Warwickshire, and the lake for sailing and angling at Pennington, Lancashire.

During the 1960s several large schemes of reclamation were embarked upon in the coalfields, notably with the aid of the Opencast Executive of the NCB, which was often able to combine the extraction of coal with the restoration of existing dereliction. R. T. Arguile has described several such schemes in the East Midlands (1971), the principal results of which are summarised in Table 17. In addition to the 80ha or so of derelict land included in these schemes of opencast coal working, the Derbyshire CC had projects in being, or actively planned, to reclaim a further 400ha of dereliction caused by coal mining from 1968 onwards. One such project at Pilsley colliery is illustrated in the plates on pp 49 and 86. The bulk of the Derbyshire reclamation schemes were designed to provide amenity and agricultural land, as only four out of a total of eighteen sites had any provision for industrial development and none was to be used for housing.

The English examples cited above form part of a nationwide trend to improve the environment of the coalfields, which is also to be seen in Wales (see pp 251–8) and, after a belated start, in Scotland. In its evidence to the enquiry into land resource used in Scotland (1972) the NCB revealed that it had disposed of 120ha of colliery spoil tips to local authorities for reclamation purposes since 1967, and proposed to release a further 220ha by 1976. In addition to disposing of 26 million tons of shale in these tips the NCB was also marketing 3 million tons of minestone per annum at active collieries, though, in spite of this, it estimated that it would need a further 15ha of

TABLE 17 Derelict land reclamation at four sites in Derbyshire

Site (collieries)	*Bonds Yard (Bonds Main)*	*Parkhouse (Parkhouse)*	*Steadmill (Alfreton)*	*Shipley Lake (Coppice, Woodside)*
Extent*	24·7ha	78·6ha	193·2ha	197·6ha
Opencast coal extracted	95,500 tons	300,620 tons	965,200 tons	1,156,700 tons
Volume of spoil tips	304,000m³	785,650m³	1,064,900m³	2,295,000m³
Other forms of dereliction	*Railways, colliery yard, shafts*	*Railways, ponds, shafts, colliery yard*	*Sewage works, subsidence lakes, shafts*	*Railways, canal feeder, colliery yards*
Expected costs of reclamation (£ per ha)	2,720	2,965	2,720	2,840†
End uses of site	—— principally agricultural land		——————	country park

* Including open cast workings on non-derelict land; derelict land accounted for about 16 per cent of the total area

† Reduction of tip and landscaping only

Data from R. T. Arguile (1971)

land per annum on which to tip colliery spoil. The major reclamation project now proceeding in Scotland is at Lochore Meadows, Fife, where 900ha of abandoned collieries, spoil tips and subsidence lakes will ultimately be reclaimed to provide industrial sites, agricultural and afforested land, and a country park. This is one of several schemes located on the declining Fife coalfield (McNeill 1973). There were also proposals to reclaim 105ha of derelict colliery land on the approaches to Edinburgh in East Lothian during the period 1968–73, although as with the parallel proposals to remove many of the oil-shale bings to the west of the city, little appeared to have been achieved by the spring of 1973.

The scale of reclamation in many coalfields has, however, been dramatic since the mid-1960s. Before that date most attention centred on schemes of reclamation which did not require major civil engineering works, for, as J. R. Oxenham had observed in 1952, 'no large-scale tools and methods have yet been used on restoration work in this country'. In the 1950s most authorities seemed to agree that afforestation offered the best prospects, and the Town Planning Institute's memorandum on industrial waste disposal (1950) considered tree planting to be the only suitable mode of treatment for colliery spoil heaps. These ideas may well have been influenced by experience in the Ruhr, where the regional planning authority (SVR) had embarked on its programme of 'turning the Ruhr green' in 1926. The policy of afforesting spoil heaps was reactivated by the SVR after 1945, and between 1951 and 1968 the authority planted 453ha of spoil heaps, together with a further 800ha of plantations, some of which were on colliery sites and on land damaged by subsidence (SVR 1970a). The authority's plans for future development include further tree planting, as well as using reclaimed colliery land for industrial building and other urban uses (SVR 1970b). Without doubt the existence of the SVR has provided the most comprehensive basis in the world for integrated land-use planning, including the reclamation of derelict land, and this contrasts very sharply with the fragmentation of planning in the equivalent areas of Great Britain, which may not have been entirely resolved by the reorganisation of local government in 1974.

Elsewhere in Europe dereliction in coal-mining districts has been the subject of reclamation projects during and since the 1960s. In Belgium legislation devised to enforce the reclamation of abandoned collieries had existed since 1911, but it was not until 1962 that steps were taken to relate colliery reclamation to regional planning policies. In particular the use of colliery spoil for road construction was encouraged and cleared sites were afforested or used for amenity purposes. Even so there were still 1,925ha of derelict colliery spoil heaps in the Sambre-Meuse coalfield alone in 1970. The Upper Silesian industrial region provides a further example of dereliction largely related to coal mining but of an unusual kind. Within the coalfield there were 1,500ha of derelict waste tips in 1970, much of it at collieries but some also at metal smelters and power stations. But the largest area of disruption was provided by the sand pits worked to provide the fill used in the podsadzka method of coal mining (see p 135). These workings covered an area of 5,000ha, of which roughly 25 per cent was derelict. Polish law places an obligation on mineral operators to restore damaged land and, during the 1960s, reclamation progressed at the rate of 300ha per annum in the Katowice region.

RECLAMATION OF SURFACE MINERAL WORKINGS

Surface mineral workings have the lowest rates of actual and potential reclamation in Great Britain, principally because deep excavations present particularly intractable problems of reclamation, notably in those cases where the workings are situated in remote rural locations. A subsidiary reason for their low reclamation potential is that some local authorities do not include flooded gravel workings and the like as derelict land justifying treatment, even though the sites may ultimately be put to beneficial use. Thus in the East Riding of Yorkshire the area of derelict excavations and pits fell by 65 per cent in 1969–71 because 180ha of derelict gravel workings in Holderness RD were judged either to be capable of further extraction or already actively used for recreation. In Kesteven a reduction of 66 per cent in the area of derelict pits and excavations during 1970–71 was attributed to reappraisal which showed that 'many of the pits were blending into the countryside with natural vegetation and wildfowl becoming established'. The reclamation of surface mineral workings has formed the subject of several papers by S. H. Beaver (1955, 1960, 1961) who, among other things, has outlined the problems of restoration according to the type of working. Generally speaking, shallow workings (less than 6m deep) have caused fewest problems, while deep quarries with low ratios of overburden to extracted mineral have caused most. In addition to the physical configuration of the abandoned workings their location in relation to sources of filling material, and the existence or absence of planning conditions enforcing restoration, have also been important factors helping to explain the pattern of reclamation. In the case studies which follow some of the major types of reclamation scheme are outlined.

Reclamation of shallow dry surface workings

Shallow pits left dry on abandonment present few technical problems of reclamation and rarely remain derelict for long under modern conditions of mineral working. However, derelict workings of this type remain from earlier periods of mineral extraction, notably in the form of opencast ironstone patches along the north crop of the South Wales coalfield. These areas of severely gullied abandoned workings have been the subject of several reclamation schemes, notably those carried out in conjunction with modern opencast coal working.

The Tafarnau Bach, Tredegar, restoration project illustrates the potential offered by this kind of derelict land, much of which provides excellent prospects for industrial development along the line of the Heads of the Valleys road. The site comprises 70ha, of which 20ha is not derelict but essential to the full use of the reclaimed land. The abandoned workings constitute an area of low hummocks and shallow hollows, with no risk of further subsidence as there was no later mining beneath the site. The main object of the reclamation scheme is to provide sites for factory building, which will be achieved by producing nine plateaux of varying size, giving a total level area of 46ha. The remainder of the site will be landscaped and the surface water drainage channelled into an artificial lake which is being created on the adjacent Bryn Bach reclamation scheme. The total cost of the scheme is estimated to be £370,239, but only the derelict part of the site qualifies for state aid; the local authority will have to fund a loan of at least £95,000 over a period of 60 years, illustrating a major difficulty encountered in reclaiming land interspersed with non-derelict areas that are to be incorporated in the restoration project.

Most commonly reclamation of shallow dry workings occurs at the sites of sand and gravel pits, many of which are restored immediately after abandonment. In rural areas most of the worked-out pits are restored to agriculture, in many instances being intensively cropped within a short period after reclamation (see plate, p 121). In urban areas pits are more likely to be reclaimed as sites for industry or housing (see plate, p 121), often following a programme of controlled refuse tipping. Consequently only those dry gravel workings which reach exceptional depths are likely to pose problems of reclamation. For example, the Singret gravel pits north of Wrexham work to a depth of 35m, and although planning consents for this locality contain restoration clauses it has not been possible to introduce progressive reclamation as commonly practised at shallower workings. Approximately 200ha of semi-derelict gravel workings were recorded in Wrexham RD in 1972 for this reason (Jacobs 1972).

Reclamation of deep dry workings

In assessing the relative difficulties of reclaiming different types of dry workings S. H. Beaver (*loc cit*) drew a distinction between excavations working thin seams (eg of coal and ironstone) and those working both stratified and unstratified minerals of greater thickness (eg limestone and china clay). Although the differences in total depth of these two types of working are no longer as great as they were in the 1950s, the distinction is still a useful one on other grounds. The excavations working thin mineral seams normally have short lives, and the high ratio of overburden to mineral extracted ensures that there are adequate filling materials for reclamation. The excavations working thick mineral deposits have much longer lives, the ratio of overburden to mineral extracted is normally very low, and the mode of extraction rarely makes it possible to reclaim part of the site while the remainder is still active. It is not surprising, therefore, that very little reclamation of this last-named type of deep working has been carried out, and that most progress at such sites in recent years has centred on screening plant and waste heaps. For example, research is being carried out to discover means of vegetating the waste dumps at china clay workings in Cornwall and Devon. However, in areas of high amenity value more ambitious reclamation projects have been undertaken. At Dove Holes, on the margins of the Peak District National Park, a scheme to restore a particularly conspicuous area of derelict limestone quarries and spoil heaps covering 22ha was initiated in 1973.

Reclamation of coal and ironstone workings

As noted above, the statutory obligation to restore opencast coal sites since their reintroduction in 1942 has led to the development of advanced techniques of reclamation which have increasingly been applied both to the treatment of derelict land and to the provision of amenities after coaling has ceased. Thus in Northumberland part of the Acorn Bank site was restored to provide a golf course (at a saving of £140 per hectare on the normal costs of restoration to agriculture), and in the same county the Druridge Bay country park has been designed to incorporate the reclaimed workings of the Coldrife opencast site. During the period 1951–70 the contractors employed by the NCB restored approximately 18,000ha of worked-out opencast coal mines. Although none of this land had ever become technically derelict at the end of coaling, it can be seen that the policy of automatic reclamation has had considerable beneficial effects in the coalfields. In the East Midlands a similar policy of 'instant reclamation' has been enforced at ironstone workings since 1951. Here the mechanics of administering reclamation are unique in Great Britain, in that the cost is borne partly by the state and partly by a levy on mineral operators. The mineral operator is obliged to bear the first £275 per hectare of reclamation costs (fixed as a norm in 1951), and the balance comes from a restoration fund based on a tonnage levy of 1.25p per ton of ironstone raised, 75 per cent being met by the operator and 25 per cent by the state. In the years 1951–71 2,360ha of ironstone workings were restored at a cost to the fund of £725 per hectare. Much of the land is immediately restored to agriculture (see plate, p 50) and in addition much former hill and dale left by earlier workings has also been restored (see plates, p 104). It has also proved possible to build on reclaimed ironstone workings, providing that sufficient time is left for the ground to settle, and this may be of crucial importance in relating ironstone quarrying to the expansion of Corby New Town.

Opencast coal working outside Great Britain has also produced outstanding developments in the techniques and management of reclamation schemes. A major European example is provided by the careful integration of extraction and restoration in the Cologne brown-coal field (see p 179), where almost all the land has either been restored to agriculture or afforested, and a considerable literature on the planting of restored opencast workings has resulted from this experience (Knabe 1964). In East Germany extensive reclamation of brown-coal workings left derelict before 1945 has taken place, and currently all the worked areas are restored after use. Many of the most recent projects have concentrated on the provision of recreational facilities, including a 300ha artificial lake, the Knappensee (50km north of Dresden), and a proposed 900ha of inland water devoted to various amenity uses at Senftenberg in the same locality.

In the United States the enforcement of restoration after cessation of strip mining has been far less rigorous, but there has been considerable research into the problems of reclamation, as is shown by the bibliographies on the topic (see p 321). As in other parts of the world, many of the most recent and most impressive projects have been designed to provide recreational amenities. Thus in Ohio, the Sallie Buffalo Park, reclaimed from abandoned strip mines by the Hanna Coal Company, covers 170ha and provides lakes for water sports, picnic areas and forested plots. American experience suggests that informal recreation areas of this type, capable of use for a longer period each year by a greater cross-section of the population than conventional sports fields, are particularly well suited to the diverse topography of abandoned mineral workings.

Whether developments on this scale are generally possible in Great Britain is open to doubt, although it is clear that some of the country parks being created from areas of dereliction take their inspiration from such projects.

Reclamation in the Oxford Clay vale

The Oxford Clay vale between Peterborough and Bletchley has the largest concentration of major brickworks in Great Britain. The Peterborough area is the scene of a major reclamation project, initiated in 1963, whereby PFA brought 80km by rail from power stations in the Trent valley is dumped into the flooded clay pits and subsequently restored. The project, devised by the CEGB, was intended to reclaim approximately 1,200ha of derelict land over a period of 30 years or so. By 1973 only about 40ha had been reclaimed, and it was being suggested that the successful marketing of PFA for other purposes would retard the progress of the project by the late 1970s. While it may be argued that the substitution of PFA as a constructional material reduces the need to work natural mineral occurrences, it is equally clear that this policy reduces the amount of filling material available for reclamation schemes. Ironically, much of the PFA is used as an aggregate and therefore diminishes the demand for sand and gravel worked by an industry which has become increasingly conscious of the need to reclaim its pits and quarries; and the argument that the commercial use of PFA as a constructional material saves additional land from dereliction may come to have little validity. In addition to providing dumping grounds for PFA the major brick producer in the Clay vale, the London Brick Company (LBC), has also provided facilities for tipping towns' refuse into its abandoned pits. In 1970 the LBC formed a subsidiary company 'to promote and develop the use of worked-out clay pits in a commercial and practical manner'. Some have questioned the morality of making a profit from self-inflicted dereliction (Barr 1969), but a more pertinent question may be whether the normal commercial concepts of profitability will permit a rapid and wholly satisfactory solution to the land-use problems of the Clay vale.

In Bedfordshire approximately 560ha of clay pits lay derelict in the late 1960s (see pp 168–70), and the history of reclamation proposals since that time illustrates the complexity of finding an adequate solution to the problems of filling deep workings even in a relatively accessible locality. Initially it was proposed to fill the pits with towns' refuse brought by rail from London, but this was rejected on the grounds that is was cheaper to incinerate waste in the city than to transport it to Bedfordshire. A second proposal, to bring colliery waste by rail from Nottinghamshire, was actively investigated, but by 1972 this too appeared to be impracticable (Cowley 1972): no alternative source of filling material appears to be available. In spite of the existence of stringent planning conditions, disappointingly little reclamation of derelict land took place between 1952 and 1970. Only 117ha were restored, equivalent to about one-fifth of the area being worked in 1970. In 1971 a further 77ha of land were reclaimed when Stewartby Lake (see plate, p 67), a flooded claypit worked before 1947, was leased by LBC to the local authority as the nucleus of a country park. It seems likely that the absence of solid waste will make it necessary to adopt a new reclamation strategy, involving the more even spreading of callow waste over the pit floors or its use over a more limited area as a solid filling, in both instances coupled with carefully controlled flooding for water storage and amenity use. The present method of tipping callow in irregular mounds on the pit floor (see plate, p 67) is inimical to restoration if imported fill is not available, and the cessation of this practice would appear to be a small burden to place on the LBC but a major contribution to the future restoration of the claypits.

Reclamation of wet mineral workings

The principal form of wet mineral working is for sand and gravel. For many years abandoned wet pits seemed to present intractable problems of dereliction, and infilling was assumed to be the only satisfactory form of reclamation. The most successful restoration schemes were located in Middlesex, where 600ha of wet pits were filled between 1918 and 1953, but even so, 640ha of derelict lagoons survived in 1954 (Beaver 1960) in a locality which could provide large quantities of solid waste and substantial demand for the reclaimed land. The most suitable filling materials were those with no organic content, but this precluded the most readily available source of fill – towns' refuse. During the early 1950s experiments in the controlled tipping of towns' refuse into wet pits were carried out at Egham (Furness 1954, Ripley 1960), but although successful, the method was not widely copied. In some areas, however, a new source of solid fill was provided in the 1950s by PFA from power stations. This had the advantage that it could be pumped as slurry to the point of disposal and could subsequently be restored for agriculture by various techniques of soilless treatment (Barber 1972). Schemes of reclamation using this medium were adopted at Brimsdown, Middlesex, and in the Trent valley (James *et al* 1961), and new projects are still being initiated. Thus in 1972 it was announced that PFA slurry was to be pumped 13km from Cottam power station in order to reclaim 160ha of derelict gravel workings near East Retford, Nottinghamshire. Unfortunately this project also reveals the difficulties of long-term reclamation planning, for within a year of its inception serious problems of relating current gravel production to phased restoration had arisen.

During the 1960s it had also become evident that it was no longer necessary to infill wet pits as a means of reclamation. Wet sand and gravel pits became amenities if left as lagoons, subject to their banks being suitably restored. Many became important centres of inland sailing and cruising (eg Billings Aquadrome near Northampton), while others were developed commercially for angling (eg Meriden, Warwickshire). Demand for such facilities was particularly strong in southern and midland England (Patmore 1970), where it is proposed to set up some of the largest 'water parks' based on former gravel workings. The Cotswold Water Park in the upper Thames valley may ultimately provide 1,600ha of artificial lakes for various water sports and wildlife sanctuaries. Former wet gravel pits also figure prominently in the planned regional parks designated for the London area: Thorpe Water Park, Staines; Colne Valley Park; and Lea Valley Park. Indeed it is argued in the case of the last-named that post-war infilling of wet gravel pits has robbed the park of important areas of water. The point has now been reached where gravel workings are planned to meet future recreational needs. Thus the national water sports centre at Holme Pierrepoint, Nottingham, originally located on the site of wet gravel pits, is to be extended by phased extraction of gravel and restoration of the pits during the next 20 years or so. One major company, Ready Mixed Concrete Ltd, operates a wholly-owned subsidiary whose task it is to develop wet pits for amenity purposes, thereby indicating the importance attached to this form of reclamation. On the other hand there are those who argue that the creation of these man-made lakes may outstrip recreational demand, and that it will continue to be necessary to find inert filling materials in order to restore abandoned workings (Jones 1973).

Reclamation of other forms of derelict land

The principal topic to be considered here is the reclamation of derelict railways and canals, but before doing so brief mention should be made of the major problems posed by dereliction caused by public utilities, especially the disposal of PFA. The disposal of PFA first became a major problem and source of potential dereliction during the early 1950s (p 195). Three possible solutions were investigated by the CEGB: the use of PFA as solid fill for various types of reclamation scheme, its reclamation *in situ* at power stations, and its use as a raw material in the construction industry. The success of the last-named course of action has reduced the significance of the other two, but imbalance between current areas of production and market demand means that some areas still yield a surplus of waste for disposal, and in 1971–2 4·2 million tons of PFA needed to be dumped prior to reclamation. In almost every instance the restored waste tips are put to agricultural use, and this is readily achieved at low cost. The development of techniques of reclamation designed to cope with the growing output of PFA during the 1950s may be considered one of the major triumphs of post-war advances in dealing with the problem of derelict land.

Reclamation of derelict railways and canals

Derelict railways present a variety of problems of reclamation which reflect the physical characteristics of the abandoned lines, their locations and the legal sanctions which may still apply to them. It is impossible to be precise about the amount of derelict railway land which has been reclaimed, because before 1974 local authorities were not obliged to record its restoration separately, and many have placed a rather open construction on what constitutes the reclamation of derelict railways in the past. In his valuable report on disused railways in the countryside J. H. Appleton (1970) indicated that 33 per cent of the route mileage of abandoned railways in England and Wales had been sold by British Rail between 1948 and 1968, and in Scotland the proportion sold between 1948 and 1971 was 38 per cent (Parham 1973). In many instances planning authorities consider this sale of land as reclamation, irrespective of what happens to the track formation after the transaction has taken place.

Railway reclamation falls into two categories, agricultural and non-agricultural. For obvious reasons the former is commonplace in rural areas where, if local authorities do not wish to buy derelict railway land, first refusal is given to trackside landowners. This proviso does not extend to stations and yards, which may command a higher price in the open market, but much of the track formation has been absorbed into adjacent farm land and has then been reclaimed at no cost to the Exchequer. The conversion of track formation to grassland or arable presents few technical problems, and in many cases the acquisition of railway land has also provided valuable means of access to farm property. Non-agricultural uses have been less common in the countryside, and some of them have been unwelcome to the farming community, where they have increased the risks of trespass and attendant vandalism. This applies particularly to the conversion of railways into 'recreation routes' (ie footpaths – commonly termed 'walkways' by planners with no apparent good reason – bridleways and cycle tracks), which have been created in several parts of the country.

The development of recreation routes is particularly well suited to abandoned railways in scenically attractive areas, or to those which can be used to link urban areas with the surrounding countryside. Actual or proposed routes such as the Tissington Trail in Derbyshire, the Derwent Walk in County Durham, and the Wadebridge–Padstow path in Cornwall, provide examples of imaginative schemes of railway reclamation, while the city of Stoke on Trent's 'greenway strategy' applies similar methods in an urban setting. This reclamation project largely makes use of abandoned mineral railways, together with much of the Potteries loop line, which was closed to traffic in the early 1960s. The system of 'greenways' (paths) will ultimately link a series of reclaimed derelict sites, providing a recreational route between the inner parts of the city and its rural outskirts. The use of abandoned mineral lines, linking industrialsi tes which are themselves derelict, provides a major opportunity to relate the greenways to other amenity schemes devised as a means of restoring derelict collieries and other types of mineral working.

The Wirral Country Park in Cheshire provides another example of the use of an abandoned section of railway as the focus for amenity provision accessible to major urban areas. In neither case has the reclamation been as inexpensive as conversion to agricultural uses in rural areas. The Wirral Country Park cost approximately £20,000 per route kilometre and the Stoke on Trent greenways are likely to have cost about half this amount, although strictly comparable data are not available. Unfortunately many potentially fine recreation routes have been ruined by the practice of dismantling bridges and selling sections of the route piecemeal before the possibilities of systematic restoration had been canvassed. Among other things this means that sections of abandoned railway remain derelict in scenically attractive areas, whereas with a little forethought they could have been incorporated into major amenity improvements.

The reclamation of canals presents similar challenges and opportunities to the reclamation of railways. In urban areas there has long been a temptation to infill abandoned canals, if only to remove a potentially dangerous nuisance. In recent years, however, canals have become increasingly important as recreational amenities, and there have been several reclamation projects which recognise this fact. Important sections of inland waterway have been reopened to cruising traffic, as, for example, the Stratford on Avon canal, and derelict canal basins have been refurbished to serve as moorings, as at the Cambrian Wharf in Birmingham (Langley-Smith 1970). Elsewhere major reclamation schemes are in being or proposed for the Peak Forest and the Cromford canals, which will ultimately form part of a series of recreation routes based on former railways and commercial canals in and around the Peak District National Park.

It has not always proved practicable to restore canals as waterways for pleasure cruising. In some instances the infilling of canals has been an unfortunate short-term expedient, reducing the variety of potential cruising routes. In others it has seemed unlikely that the restoration of a canal to cruising standards would be worthwhile, and in these cases alternative forms of reclamation have been adopted. Thus parts of the Rochdale canal in Manchester have been reclaimed to provide a linear park, and it is proposed to extend this and to carry out similar treatment on the Stockport branch of the Ashton canal. Reclamation work of this kind is, like its railway equivalents in urban areas, exceedingly expensive. The expenditure on the Rochdale Canal Park amounted to approximately £112,000 per kilometre, notwithstanding the fact that there were no costs of acquisition. The cost of the proposed extension to the park, covering an area of 4·5ha, was estimated as £192,000, of which about two-thirds was for acquisition of the land. Thus the conversion of the canal into a much-needed recreational amenity serving a densely populated sector of the city cost approximately £700,000. There can be no doubt that reclamation of derelict railways and canals for amenity purposes is an expensive exercise, although the benefits which such projects bring to deprived urban areas cannot readily be quantified on a balance sheet.

RECLAMATION OF DERELICT LAND IN SOUTH WALES, 1967–72

South Wales provides a striking example of the progress in reclaiming derelict land since the legislative changes of 1966 in a locality which had a severe burden of dereliction. In 1964 T. M. Thomas identified 6,800ha of derelict land in South Wales, excluding derelict railway land, of which 46 per cent was attributable to coal mining and 32 per cent to primitive forms of opencast ironstone working (Table 18).

TABLE 18 Composition of derelict land in South Wales, 1964

	per cent		*per cent*	*ha*
Colliery spoil	41			
Derelict collieries	3	Coal	46	3,128
Mining subsidence	2			
Ironstone patches	32	Surface minerals	42	2,856
Quarries	10			
Industrial sites	4	Other industry	12	816
Metal smelting slag	8			

Data from T. M. Thomas (1964)

The highest densities of derelict land (over 100ha per 1,000) were situated along the north rim of the coalfield between Merthyr Tydfil and Blaenavon, with a less dense concentration (between 50–100ha per 1,000) along the eastern outcrop in Monmouthshire and in the Rhondda valleys. The amount of reclamation carried out before 1964 was very slight, amounting only to about 130ha for housing, factories and recreational areas (including land cleared for industrial development at Merthyr Tydfil in 1937–9) and 200ha afforested or in the ownership of the Forestry Commission. Post-war progress had been particularly disappointing, largely because of financial restrictions imposed by the central government during 1952–9 and the difficulty of planning long-term reclamation schemes under the form of grant structure operating during 1960–65. In 1966 a Derelict Land Unit was established at the Welsh Office, and it is from this organisation that the data analysed below have come. The programme of reclamation is administered by local authorities, and the study of Merthyr Tydfil derives from the work of the Monmouthshire Joint Committee, whose reports on derelict land would serve as a model for a national inventory of dereliction, were one ever to be compiled. (Although geographically in Glamorganshire, Merthyr Tydfil CB joined the committee in 1972.) In addition the lower Swansea valley was the subject of a particularly detailed survey of derelict land published in 1967.

The pattern of reclamation shown in Fig 32 relates to grant-aided schemes approved during the period 1967–72. Not surprisingly the distribution of reclamation projects generally reflects the pattern of dereliction revealed by the 1964 survey, except that little work had been carried out in Merthyr Tydfil and along the southern margins of the coalfield in the Maesteg-Ogmore district. The major areas of reclamation lay in northern Monmouthshire, where most of the dereliction stemmed from deep coal mining and opencast ironstone working, and the Swansea valley, where most of the derelict sites were abandoned metal-smelting plants and their slag heaps. Table 19 shows the general structure of reclamation in South Wales during 1967–72.

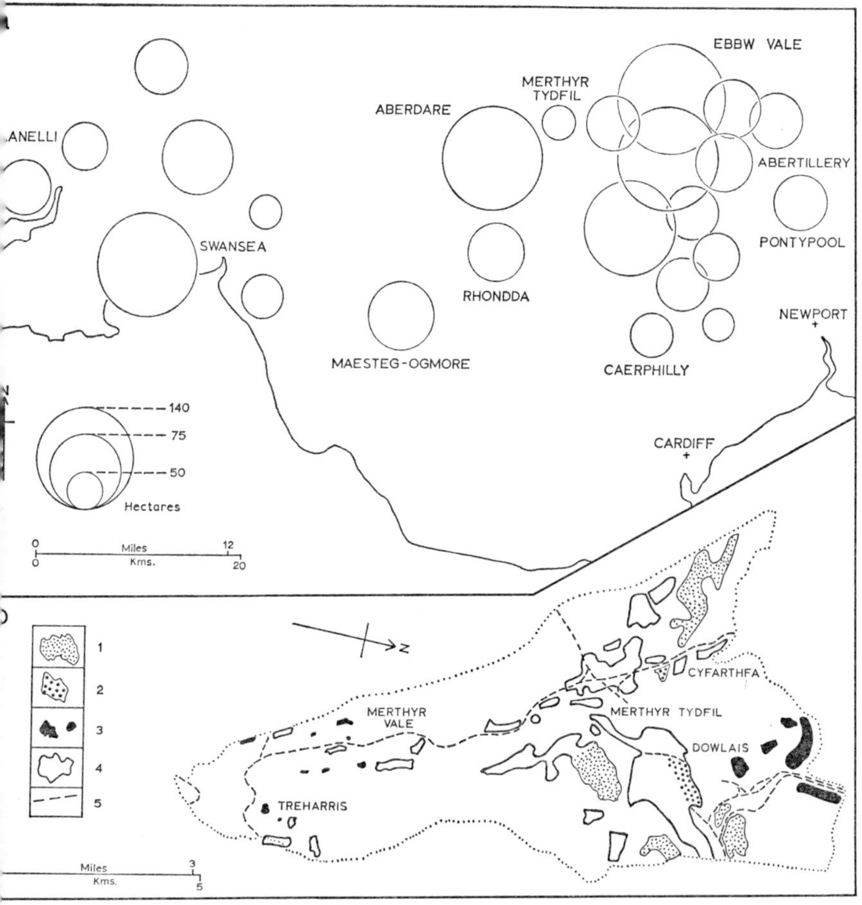

Fig 32 Reclamation of derelict land in South Wales (a) Reclamation schemes receiving grants in aid, 1967–72, by local authority areas. Returns for some areas have been merged for sake of clarity: schemes of less than 5ha not shown; (b) Derelict land to be reclaimed in Merthyr Tydfil CB: 1, ironstone patches; 2, ironworks slag; 3, quarries; 4, other derelict land, mainly colliery waste; 5, derelict railways. Based on a survey carried out in 1972 by the Monmouthshire Joint Committee. For a map of dereliction in Merthyr Tydfil in 1964 see T. M. Thomas (1966), 131

TABLE 19 Derelict land reclamation in South Wales, 1967–72

Origins	*Extent (ha)*	*Percentage of total*	*Cost (£000)*	*Percentage of total*	*Costs per ha (£)*
Coal mining	621	37·4	3,508	42·1	5,649
Transport	87	5·2	462	5·5	5,310
Metals industry[1]	129	8·0	1,059	12·7	8,209
Other industry	38	2·3	164	2·0	4,316
Surface mineral working	27	1·4	168	2·0	6,222
Other and mixed[2]	759	45·7	2,974	35·7	3,918
TOTAL	1,661	100·0	8,335	100·0	5,018

[1] Iron and steel and tinplate works; non-ferrous metal refineries and slag

[2] Mainly coal mining, iron smelting and surface mineral working complexes along north crop

Data are from returns made by the Welsh Office and relate to approved grant-aided schemes

It is one of the weaknesses of the data that almost half the reclaimed land has to be placed in the category 'others' and 'mixed types of dereliction', although much of this is a genuine mixture of ironstone patches, colliery spoil heaps, abandoned buildings and railways such as might be expected in defunct mining areas. Land in this category proved cheapest to reclaim, largely because so much of it was low-amplitude dereliction requiring minimal attention. Dereliction caused by coal mining produced the next largest area of reclaimed land, at the high cost of £5,649 per hectare. Some of the schemes involved the removal of massive waste heaps: the six largest, in terms of area, are listed in Table 20, which reveals the great variation in costs per hectare in this category. The most expensive reclamation schemes, however, were those relating to abandoned metal-smelting plants in the Swansea valley, where the average cost of reclamation was £8,209 per hectare. As Table 21 shows the gross costs of reclamation schemes in this category also varied appreciably and, even where the reclaimed land had a recognised after-value (indicated by the difference between gross and net costs), the net costs of restoration were still higher than the average for colliery land.

It is instructive to compare the estimates of the costs of reclamation made in 1966 by the Lower Swansea Valley Survey (Hilton 1967) with the actual costs incurred since that date.

TABLE 20 Examples of reclamation costs at collieries in South Wales

Colliery site	*Extent (ha)*	*Gross cost*	*Gross cost per ha*
		£	£
Gilfach Goch group	50	196,096	3,922
Nine Mile Point	37	198,023	5,352
Ebbw Vale 7/8 tips	35	83,140	2,375
Ebbw Vale Marine	32	174,122	5,441
Lewis Merthyr	20	140,744	7,037
Crosshands	24	92,910	3,871

In each instance the sites had no after value and the gross and net costs of reclamation were, therefore, identical

TABLE 21 Costs of reclamation in the Swansea valley, 1967–72

Site	*Extent (ha)*	*Gross cost*	*Gross cost per ha*	*Net cost per ha*
		£	£	£
Hafod tip	3·6	165,874	46,076	46,076
White Rocks tip	48·5	129,210	2,664	2,664
Duffryn steelworks	11·0	161,081	14,644	6,916
Worcester/Forest works	30·5	294,554	9,657	6,406
Pontardawe steelworks*	31·5	276,919	8,791	8,791

* Outside the area of the Lower Swansea Valley Project

The survey estimated that it would cost £3,077 per hectare to treat the 260ha of derelict land which it had identified. By 1972 approval had been given for the reclamation of 228ha, at an average gross cost of £4,760 per hectare. Part of the discrepancy may be attributable to recent inflation, but it is also clear that in a complex derelict area such as the lower Swansea valley some of the dereliction will cost appreciably more to reclaim than the expected average (Ward 1971).

The cost of reclaiming derelict transport installations in South Wales averaged £5,310 per hectare (Table 19). With the exception of reclamation along the line of the Swansea canal at a cost of £5,146 per hectare, most of the remaining expenditure was on railway land. A major problem stemmed from the fact that many of the abandoned railways had crossed the deeply incised valleys of the coalfield on tall viaducts. Although these structures covered small areas, they were very expensive to remove, a case in point being the Taffs Well viaduct, north of Cardiff, which covered only 0·2ha but cost £50,060 to reclaim – the highest gross cost per hectare in South Wales. The data on reclamation schemes provide no information about after use, but it is clear from the fact that very few sites had an after-value ascribed to them that the majority were not designed for immediate industrial or other economic development. The total after-value of the reclaimed land amounted to only 8·5 per cent of gross reclamation costs, and this was attributable to the reclamation of sites for industrial use in the lower Swansea valley and at a handful of locations scattered about the coalfield.

Merthyr Tydfil had the dubious distinction of recording the highest density of derelict land in any county borough in England and Wales in 1972, with a rate of 180ha per 1,000. Although Merthyr Tydfil had cleared 40ha of derelict land for industrial development on the sites of the Cyfarthfa and Dowlais ironworks before 1939, and had cleared further land for industry and housing at Pentrebach and Ynysfach respectively in the mid-1940s, very little progress had been made since that date (with the exception of work at Aberfan which was funded separately from the main reclamation programme). One reason for this was that many of the abandoned spoil and slag heaps along the north crop were intermittently worked for small coal, shale and slag, and in many instances this

Q

TABLE 22 Derelict land reclamation programme, Merthyr Tydfil CB

Type of dereliction	*Percentage*	*Derelict area 1972 (ha)*	*Percentage reclaim-able*	*Percentage develop-able*	*Area develop-able (ha)*	*Percentage developable as*				
						Industrial & urban	*Recreation*	*Agriculture*	*Forestry*	*Mixed uses*
Coal mining[1]	52·0	677	100	61·5	416·0	29	1·5	36	4	29·5
Ironstone patches	19·5	253	100	60·5	153·0	83	—	17	—	—
Quarries	6·0	74	91	9·0[2]	6·5	18	48·0	—	34	—
Ironworks slag	5·5	71	100	68·0	48·5	100	—	—	—	—
Railway land	9·0	119	44	12·6	15·0	100	—	—	—	—
Mixed sites	8·0	101	100	51·0	52·0	58	4·0	32	—	6·0
TOTAL	100·0	1,295	94	53·5	691·0	49·5	1·5	28	3	18·0

[1] Spoil tips and abandoned collieries
[2] If the Morlais Castle quarries are ultimately used as a country park, the proportion would rise to 59 per cent
Data from Joint Committee Report 1972

piecemeal extraction of waste during the post-war years intensified the problems of dereliction. The existence of such workings is frequently outside planning control, as the waste heaps may be chattels in law, and their exploitation is not 'development' as covered by planning legislation.

Dereliction in Merthyr Tydfil in 1972 is summarised in Table 22 and its pattern is depicted in Fig 32. A noteworthy feature is the great concentration of derelict land along the north crop, both along the outcrop of the coal seams and the ironstone veins and further north on the outcrop of the Carboniferous Limestone, where massive quarries were created to supply the blast furnaces of the town's ironworks. The report from which the data in Table 22 have been derived makes an important distinction between that part of the derelict area which is reclaimable (94 per cent of the total area) and that which is capable of development, including recreational uses (53·5 per cent of the total area). The lowest potential for development is shown by the quarries, although this situation would be dramatically changed were the Morlais Castle quarries to be reclaimed as part of a proposed country park. Railway land also offers little prospect for development, owing almost entirely to its linearity and unsuitability for use as recreation routes. Roughly half the developable reclaimed land is capable of attracting either housing or industry, and over a quarter may be returned to agriculture. The fact remains, however, that even in an urbanised locality which has succeeded in attracting much new industrial development since 1945, almost half the derelict land will be given simple cosmetic treatment with no form of development to follow in the foreseeable future. This is characteristic of much of the reclamation programme in South Wales, and it is perhaps one of the most heartening features of recent years that state financial aid is available for schemes which have no tangible economic benefits to offer but must inevitably improve the visual amenities of nineteenth-century industrial areas.

References to this chapter begin on p 316.

CHAPTER NINE

Derelict Land and Reclamation: Regional Case Studies

PREVIOUS chapters have concentrated on the separate elements of dereliction and reclamation, and it is, therefore, necessary to examine them together in a regional context. This also permits brief reference to the relation between dereliction and other regional land-use planning issues. Two British regions have been chosen for study: the North West and the West Midlands economic planning regions.

DERELICT LAND AND RECLAMATION IN THE NORTH WEST

In 1969 the North West economic planning region contained approximately one-eighth of the derelict land in England and Wales, at the high density of 7·3ha per 1,000 (Table 4a). In that year the greater part of south Lancashire, excluding Merseyside, had a density of dereliction in the national upper octile above

5·1ha per 1,000 (Fig 33). There were smaller tracts of land in the same category in Furness, mid-Cheshire and the High Peak of Derbyshire. The lowest densities were located in north Lancashire, Merseyside and the greater part of Cheshire. The pattern of notional costs of dereliction (as explained on p 81) was similar but less homogeneous, with notional costs above the upper octile (42 per cent) recorded in parts of south Lancashire, Furness and mid-Cheshire. However, in south Lancashire many urban areas were, by virtue of their relatively high rateable values, placed in a lower category of notional costs than of density of dereliction. The converse was true of several rural areas in north Lancashire. The major problem areas, identifiable from the conjunction of high densities and notional costs, were in the largely moribund exposed coalfield of south Lancashire, the abandoned haematite mining district of Furness, the Cheshire saltfield and the limestone quarrying area of the High Peak.

The record of reclamation in the North West is not available in a uniform manner for all local authority areas, owing to the division of responsibility for restoration projects noted below. Data are available for the economic planning sub-regions (Table 23) showing that the greatest amount of reclamation has been carried out in the south Lancashire and Manchester sub-regions, which together with the southern fringe of Rossendale cover the greater part of the exposed coalfield. The absence of reclamation schemes from Furness is noteworthy, as is the relatively poor showing of the North East Lancashire and the South Cheshire & North West Derbyshire sub-regions. The estimated gross cost of reclamation varies appreciably and, even though the highest (relating to the restoration of a clay pit in Blackpool) proved to be an overestimate, the figure for the region is still well above the national average.

In 1968 the Lancashire CC prepared its programme of reclamation for the period to 1974 in the knowledge that it was about to tackle only one-third of the recorded dereliction and that the programme might represent as little as one-eighth of

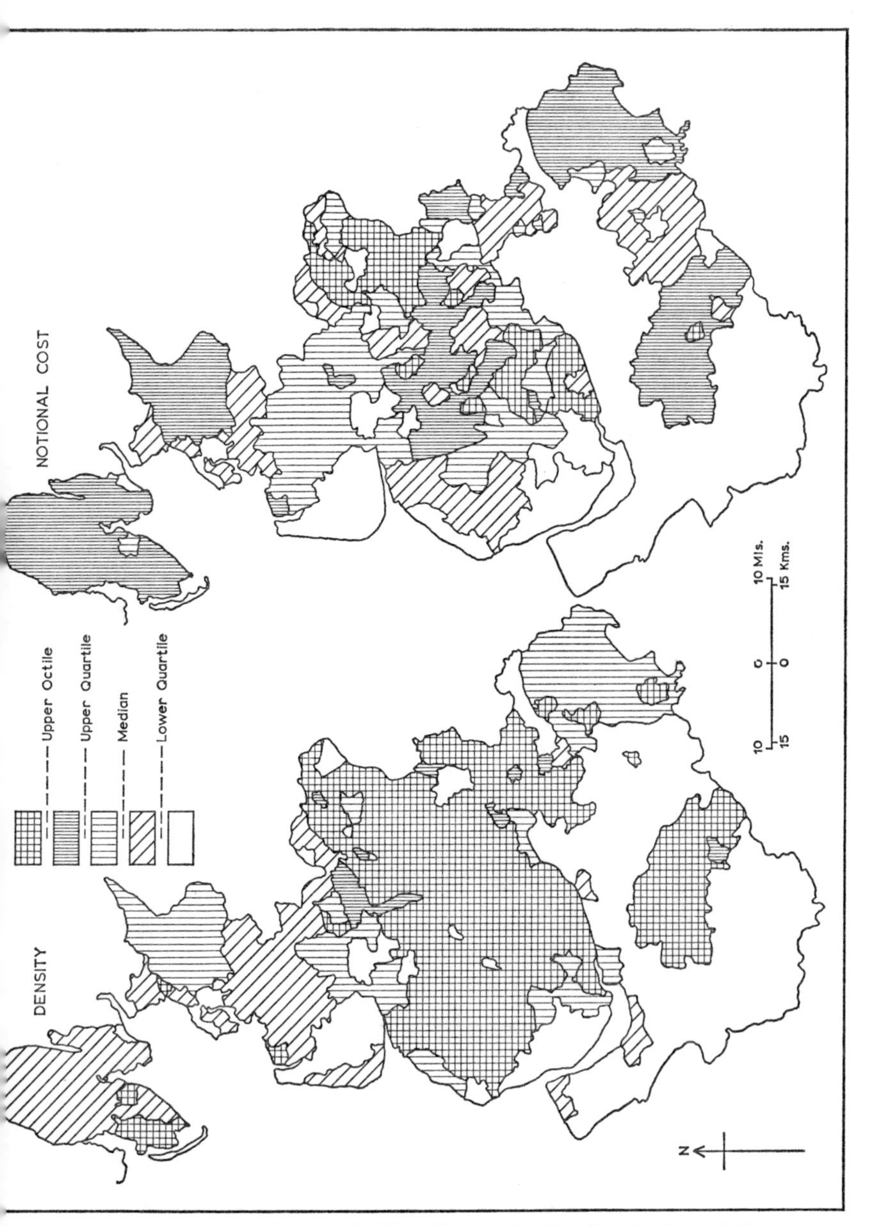

Fig 33 Derelict land in the North West Economic Planning Region, 1969, showing density and notional cost of dereliction by local authority areas. NB: in this map and Figure 38 the five categories are derived from the national values used in Figure 13. Density: upper octile 5·1, upper quartile 3·0, median 1·0, lower quartile 0·4 (hectares per 1,000). Notional cost: upper octile 42, upper quartile 21, median 8, lower quartile 1·9 (per cent). Figures 33 and 40 are strictly comparable with Figs 18-22 inclusive. Data are from returns to the Ministry of Housing and Local Government

TABLE 23 Derelict land and reclamation projects in the North West

Planning sub-region	*Derelict land justifying treatment 1967 (ha)*	*Spoil heaps per cent*	*Excavations and pits per cent*	*Other per cent*	*Programmed reclamation schemes 1967–71*		*Percentage of total area justifying treatment included in programme*
					area	*estimated cost per ha (£)*	
Furness	127	85	14	1	0	0	0
Fylde	33	0	86	14	6	12,795*	18·5
Lancaster	15	0	95	5	0	0	0
Mid-Lancashire	140	12	75	13	6	219	4·0
NE Lancashire	369	11	68	21	38	4,921	10·2
Merseyside	139	24	11	65	28	3,011	19·8
S. Lancashire	1,732	50	5	45	443	3,565	25·6
Manchester	922	37	27	36	168	5,777	17·3
S. Cheshire & NW Derbyshire	482	30	27	43	41	2,496	8·5
	3,959	39	23	38	730	4,113	18·4

Data from North West Economic Planning Council, *Derelict Land in the North West* (1969)
* This project, at Blackpool, actually cost £7,592 per ha: the accuracy of the other estimates cannot be checked

the existing and potentially derelict land, estimated at 17,400ha. The gross cost of the programme was estimated at £1,427,000, but of this approximately 96 per cent was thought to be recoverable from the disposal of land and government grants, leaving a balance of £52,000 to be funded by the local authority in the long term. In 1972 the county planning office concluded that it would take 20 years to clear the backlog of recorded dereliction, which had *increased* by 18 per cent since 1965, in spite of the progress of reclamation.

The distribution of reclamation projects completed or programmed for the period to 1976 is shown in Fig 34. Their concentration on the exposed coalfield in south Lancashire, particularly in Ashton-North Makerfield, is noteworthy. Indeed only two of the schemes completed by 1972 were located off this coalfield, although several of those projected were situated in north east Lancashire. The latter area had attracted little

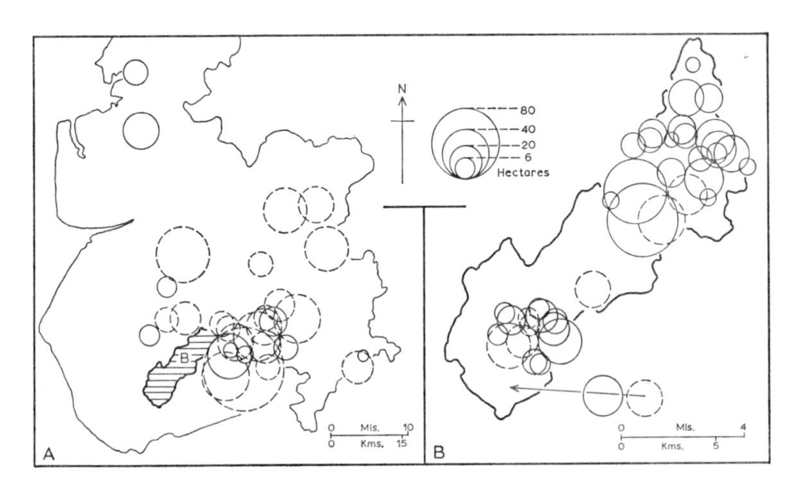

Fig 34 Reclamation of derelict land in Lancashire, 1954–76: (a) land crelaimed and reclamation programme in Lancashire AC (except North Makerfield); (b) land reclaimed and reclamation programme in North Makerfield and St Helens CB. The circles are to the same scale on both maps, and show reclamation completed or in progress, 1954–72: broken circles show the reclamation programme for 1972–6. Schemes of less than 5ha have been merged, or in the case of St Helens are shown separately by a single symbol. Data from Lancashire CC and St Helens CBC planning offices

attention, in spite of the progressive contraction of its coal-mining industry, largely because the scale of dereliction was slight. Subsidence had never been a major problem, even in the Calder lowlands, because a thickness of less than 5m of coal had been removed over most of the Burnley basin. Few large colliery spoil tips had been created because much of the coal raised was dirt-free and it had often proved possible to dump waste brought to the surface in ravines and depressions. In contrast the south Lancashire coalfield is characterised by much larger thicknesses of extraction and greater amounts of spoil. In spite of a reduction in the number of collieries in Lancashire from seventy-one in 1950 to nine in 1972, the amount of waste to be disposed of had risen by 70 per cent during the period 1946–72. The average gross cost of the Lancashire reclamation projects during the period 1954–73 was £2,416 per hectare, compared with £11,200 per hectare in the county boroughs situated within the geographical county, demonstrating the great cost difference between rural and quasi-rural schemes and those executed in large towns and cities. The costs of the county's reclamation programme are outlined In Table 24, and the contrast in expense levels before and after 1966, when more generous grants-in-aid were made available, is striking. The importance of reclamation projects to provide land for housing and industry is shown in Table 25. They formed 35 per cent of the programme completed by 1972, and their exclusion from the current programme largely reflects the nationwide trend towards reclamation for amenity purposes and agriculture, although some of the farmed land could be developed for housing and industry at a later date.

The absence of comparable data for the whole of the North West makes it impossible to discuss the progress of reclamation throughout the region in similar detail, and reflects the division of responsibility for restoration among three administrative counties and twenty-one county boroughs. It is not yet clear what effect the reorganisation of local government in April 1974 will have on the control of reclamation programmes. The

TABLE 24 Derelict land reclamation, Lancashire AC, 1954–72

	Cost per ha (£)		
Type	*1954–66*	*1967–72*	*1972 (programme)*
Industry	—	—	—
Residential/other urban	1,122	—	—
Open space/recreation	759	1,146	3,617
Agriculture	720	2,498	—
Mixed schemes*	744	4,304	1,848
TOTAL	758	3,785	2,580

Data from Lancashire CC Planning Office

* Although details are available for areas by various categories (Table 25), they are not broken down in the same way for costs

TABLE 25 Derelict land reclamation, Lancashire AC, 1954–72

	Amount (ha)		
Type	*1954–66*	*1967–72*	*1972 (programme)*
Industry	10·5	56·5	—
Residential/other urban	18·0	36·0	2·5
Open space/recreation	24·0	46·0	99·5
Agriculture	89·0	34·0	94·0
Tree planting	27·0	2·0	32·0
TOTAL	168·5	174·5	228·0

Data from Lancashire CC Planning Office

Department of the Environment has expressed the desire that existing derelict-land reclamation teams should not be disbanded, but in order to achieve this, some form of joint agency would be necessary in the North West. For example, the major problem areas within the present (1973) administrative county of Lancashire will fall within three new counties (Greater Manchester, Lancashire and Merseyside), and the remainder of the North West will be divided among Cumbria, Cheshire, Derbyshire and Lancashire. The allocation of the derelict areas outside south Lancashire to their respective new counties would present few organisational difficulties, but continued fragmentation in the south, albeit with fewer local authorities, would be to the detriment of the established restoration programme.

Derelict land in Ashton-North Makerfield

Within Lancashire the most severe dereliction is to be found in the area between St Helens and Leigh, coincident with one of the most recently abandoned sectors of the coalfield and also containing numerous forms of dereliction not related to mining. The development of derelict land in this area has been discussed by several writers (Gulley and Smith 1960, Wallwork 1960, Rimmer 1966) and this account will mainly focus on recent changes. The Ashton-North Makerfield town map areas contain the highest concentration of dereliction (North West Economic Planning Council 1969) and the largest number of reclamation schemes in Lancashire (Fig 34b). In the years 1954–73 464ha of derelict land were reclaimed (Table 26), and a further 91ha were scheduled for treatment in the 1973–6 programme, amounting to about two-thirds of the dereliction recorded in 1954. Much of the land dealt with by 1973 had been restored to agriculture or amenity use (Table 26), but provision had also been made for local authority housing (see plate, p 139) and industry. Many of the schemes completed before 1968 were inexpensive, ranging in cost from £193 to £1,600 per ha, but the provision of higher rates of grant, particularly after 1970, led to the inception of a more ambitious programme, with gross costs ranging from £628 to £4,821 per ha. The most impressive scheme, begun in 1968 and still proceeding in 1973, was that designed to remove the complex of abandoned collieries and 45m-high conical waste tips at Garswood Hall (colloquially known as 'the Wigan Alps', or more prosaically 'the three sisters') and to provide 126ha of land to be allocated to a variety of functions at a gross cost of £601,112.

An illustration of the development of derelict land in part of Ashton–North Makerfield is provided by Fig 35a and b, and the resultant landscape is depicted in the plate on p 140. Much of the dereliction stemmed from deep coal mining, but in addition

TABLE 26 Reclamation in Ashton-North Makerfield, 1954–73*

End use	*Area reclaimed (ha)*	*Percentage of total*
Agriculture	193	41·5
Education	10	2·0
Housing	30	6·5
Industry	67	14·5
Recreation	102	22·0
Tree planting	55	12·0
Other urban uses	7	1·5
Total reclaimed	464	100·0

*Ashton in Makerfield UD and the North Makerfield town map area (Abram, Aspull, Hindley, Ince in Makerfield UDs)
Data from Lancashire CC Planning Office

iron smelting and brickmaking also created derelict land. Permanent flooding due to undermining began to occur in the late 1880s, roughly coincident with the introduction of the longwall technique to the Wigan coalfield. However, it was not so much the mode of extraction as the intensity of mining which caused this subsidence, for between about 1870 and 1962, when mining ceased, approximately 21m of coal were removed from beneath the area, at depths of 45–640m. In several localities the superficial cover comprised thick peat or glacial drift, and this, combined with sluggish natural drainage that was impeded by colliery waste dumped downstream, greatly emphasised the impact of subsidence, and led to the formation of extensive lakes or 'flashes' (Fig 35a and b, and plate, p 140). In addition, increasingly large amounts of colliery waste were tipped in the area after the 1890s, particularly on its southern margin at Bamfurlong (see plate, p 122), and ironworks' slag was tipped on to Ince Moss. By the 1950s there were approximately 200ha of derelict or potentially derelict land in the locality, most of it created by mining subsidence or the disposal of colliery waste.

During the 1960s several reclamation schemes were carried out in the area at Worsley Mesnes (land reclaimed for housing and recreation), Westwood (using PFA from the adjacent power

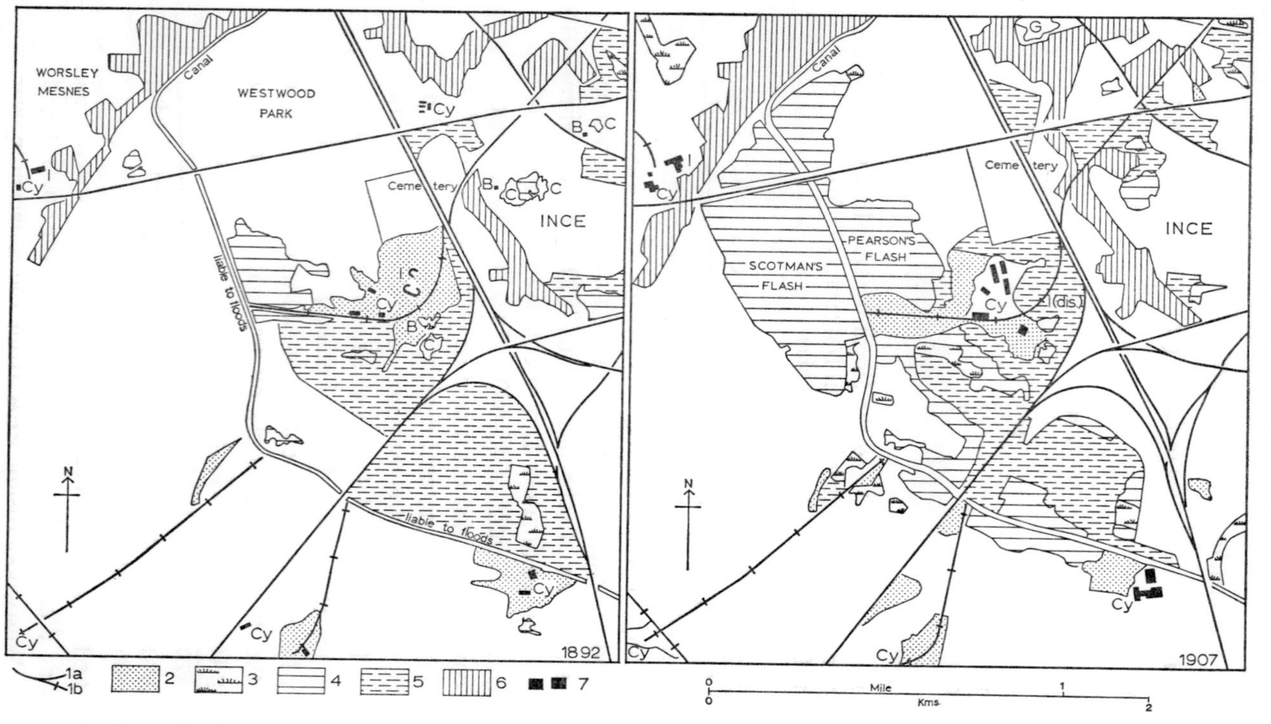

Fig 35a Industrial land use at Ince in Makerfield, Lancashire, in 1892 and 1907: 1a, railways; 1b, mineral lines; 2, ironworks, slag and colliery spoil tips; 3, marshland; 4, flooded land; 5, roughly vegetated waste land; 6, built-up area; 7, mineral working/using industrial plants (generalised); B, brickworks, C claypit; Cy, colliery; I, ironworks. For a similar map additionally illustrating the landscape in 1846 and 1927 see Wallwork (1960), 269. Data from Ordnance Survey 1:10,560 maps

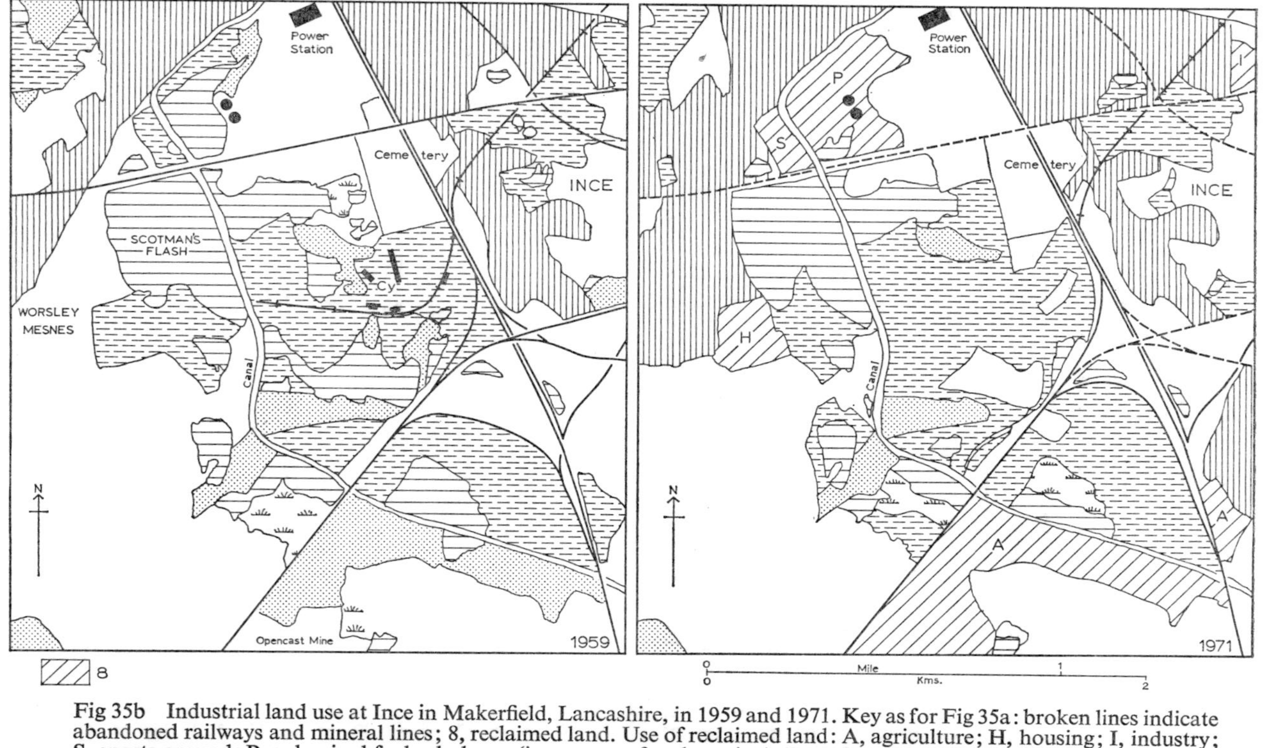

Fig 35b Industrial land use at Ince in Makerfield, Lancashire, in 1959 and 1971. Key as for Fig 35a: broken lines indicate abandoned railways and mineral lines; 8, reclaimed land. Use of reclaimed land: A, agriculture; H, housing; I, industry; S, sports ground; P, pulverised fuel ash dump (in process of reclamation). Data from aerial photography and fieldwork: see also plates, pp 122 and 140

station to fill one of the flashes) and Ince Forge (for industrial expansion). The largest scheme was at Bamfurlong (see plate, p 122), a project linked with developments at Garswood Hall immediately to the south-west. Even so, dereliction was still the dominant element in the landscape in 1971, with approximately 130ha of derelict land, together with 7·5 route km of abandoned railways.

Although the programme of reclamation in Ashton–North Makerfield has been impressive, 735ha of derelict land remained in 1971, at a conservative estimate. It is too early to say whether the twin aims of the reclamation programme as stated in 1968 have been achieved: these were 'transformation of the environment (leading to) a decrease in migration (and) an increase in private house building'.

Between 1961 and 1971 the population of Ashton–North Makerfield grew by 16 per cent, compared with a loss of 3 per cent in the previous intercensal decade. Significantly Ince in Makerfield UD, with the highest density of derelict land, was the only area to suffer a continued loss of population (12·3 per cent in all; 15·7 per cent by migration). Furthermore in the remaining local authority areas population losses were recorded in several wards which contained derelict land. Private housing estates have recently been built in the area, notably in the district to the south-west of Wigan (see plate, p 140). Here ease of access to the M6 motorway, combined with general environmental improvement and the removal of the risk of further mining subsidence, has attracted private builders to what was formerly a relatively undeveloped area. Although the threat of further mining dereliction has disappeared from the greater part of Ashton–North Makerfield, adjacent areas on the concealed coalfield to the south and east still face problems of waste tipping and of subsidence, but on a smaller scale with the advent of high-speed mining at great depth. However, much more imaginative policies of waste disposal are now being pursued by the NCB, including a programme of tipping designed to produce landforms which can be incorporated in an extensive recreational area to be located to the west of Leigh. Had similar policies prevailed elsewhere in the locality, even as recently as the 1940s, many of the restoration schemes carried out in the years since 1966 would have been easier and less expensive to mount.

Derelict land in St Helens

In 1971 St Helens CB contained 120ha of derelict land at a density of 33ha per 1,000. Although not the worst afflicted large town in south Lancashire (Bury CB, with the same area at a density of 40ha per 1,000 held this position), St Helens possesses a great variety of severe forms of dereliction and, equally important, an impressive and well documented record of reclamation. In addition to the types of dereliction commonly associated with coal mining the town contains abandoned alkali waste tips, 'sand lodges' at the glassworks, disused clay pits, and several sections of abandoned canal and railway. Fortunately the glass industry's contribution to dereliction has not been compounded by abandonment of the workings in the Shirdley Hill Sands to the north-west of the town, for these have almost always been restored after use (Taylor 1967).

The town's reclamation programme began in the 1950s with a project at the Parr industrial estate, and by 1972 187ha had been restored (Fig 34b), equivalent to 60 per cent of the dereliction recorded in 1954. Of this land 44 per cent was for industry, 17 per cent for housing, 29 per cent for recreation and the remaining 10 per cent for a variety of urban uses. The reclamation programme for 1972–7 envisages the treatment of a further 85ha, at a gross cost of £11,659 per ha. This includes the clearance of chemical waste tips which have lain derelict for upwards of 50 years, the infilling of clay pits and the restoration of abandoned railways and canals. Yet again the contrast between schemes located in a major urban area and those situated in adjacent county districts is apparent, and the high costs in St Helens are all the more striking in view of the fact that most of the restored land is for amenity uses.

Derelict land in mid-Cheshire

In terms of density and notional costs of derelict land Cheshire occupies an intermediate position between areas of severe dereliction, such as Lancashire, and those counties with very little dereliction in southern England. The major concentration of derelict land lies on the saltfield in mid-Cheshire (Fig 33), although not all of this dereliction is related to salt extraction. Sand quarrying to the east of the central ridge has in the past produced deep steep-sided pits which flooded on abandonment, but modern practice favours levelling the hummocky drift by completely removing the sand, thereby permitting immediate restoration to agriculture. In 1971 a detailed survey revealed the existence of 931ha of derelict land in Cheshire, of which 607ha were considered to conform to the official definition and to qualify for reclamation grants. Of this reduced total 31 per cent was attributable to salt working or salt-based industry, and a further 20 per cent of the county's derelict land was also located in mid-Cheshire (Fig 36). In addition about 250ha of potentially derelict land which is likely to need treatment in the near future was situated in the Northwich area, forming 75 per cent of the county's 'unofficial' dereliction. The development of derelict land and of the industries responsible for it on the saltfield has been dealt with elsewhere by the author (see references on p 312), and the major technical aspects of the mining and waste disposal problems have been outlined above (see pp 146–7 and pp 192–3 respectively). Here it is proposed to examine the principal current problems posed by dereliction related to salt working in mid-Cheshire.

Scarcely any reclamation work apart from a few cosmetic schemes, had been carried out before the late 1960s, when the adoption of a town expansion programme at Winsford encouraged the use of derelict industrial sites for factory building in the Weaver valley. At about the same time other derelict land in the valley was returned to active use for the outdoor storage of rock-salt, although the visual impact of this was similar to

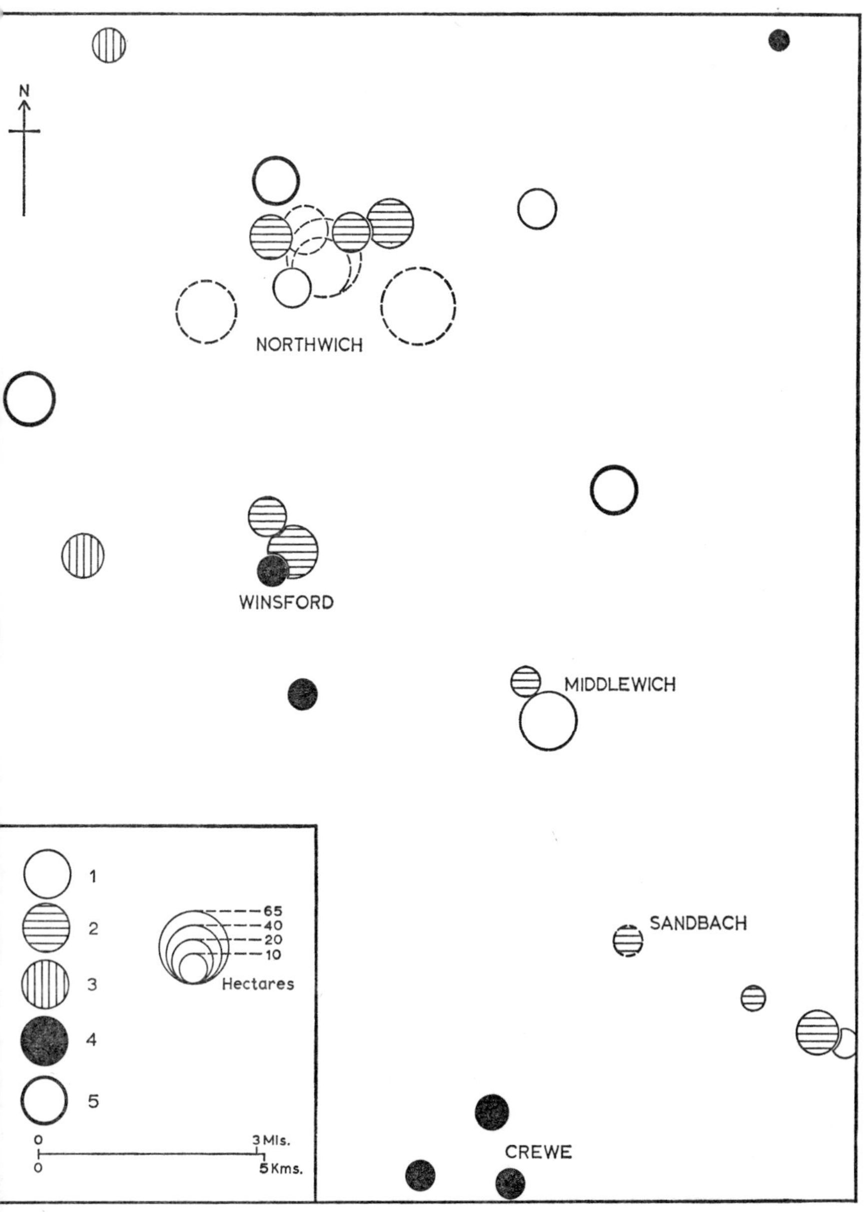

Fig 36 Derelict land in mid-Cheshire, 1971: 1, chemical waste; 2, saltworks; 3, surface mineral workings; 4, public utilities and railways; 5, military installations. Circles indicate area of derelict land in 1971; broken circles area of potentially derelict land. Data from Cheshire CC planning office and field survey

that of waste dumps! However, 55ha of derelict land survive in what was once a prosperous salt-working district (see plate, p 85), forming a scrub-covered landscape, pock-marked with subsidence hollows and littered with the ruins of brine cisterns, salt pans, warehouses and abandoned wharves and mineral lines (see plate, p 175). To the south of this devastated area Winsford Bottom Flash, formed by subsidence in the 1890s, has proved to be more amenable to beneficial use without reclamation, and as a popular sailing venue is colloquially known as 'the Cheshire Broads'. The town expansion programme is likely to ensure that much of the remaining dereliction will shortly disappear, and in addition features such as Bottom Flash and the footpath following the line of a disused railway to link with existing amenity areas in Delamere Forest are likely to be further improved. The town has the advantage of no longer being a centre of brine pumping, so that the risk of future damage from subsidence is minimal. An entirely different situation obtains in the southern part of the saltfield, between Middlewich and Sandbach, where, in addition to the sites of abandoned salt and chemical works, some of which have lain derelict for over 40 years, dereliction is still being created by natural-brine pumping subsidence. It is anticipated that natural-brine pumping will cease in this locality during 1973–4, and reclamation would then be able to proceed without fear of further damage.

The main complex of existing and potential dereliction is situated to the north of Northwich, where there are about 350ha of land damaged by salt working and dumping of chemical waste. The chemical waste dumps constitute a particularly difficult problem, particularly where they consist of large settling ponds (lime beds) constructed on sound agricultural land (see plate, p 85). Although the high retaining walls have been seeded and belts of trees have been planted at their bases, the complete restoration of the lime beds would be expensive and technically difficult. As they are no longer in regular operational use, this method of waste disposal having been replaced by stowage in the abandoned cavities left below ground by the controlled brine-pumping method at Holford, it is possible that an attempt to reclaim the lime beds will be made in the near future. It is hoped that the venture would be jointly financed by the chemical industry and government grants, but such a policy runs counter to the normal assumption that derelict land must be in public ownership to qualify for state aid.

A similar legal problem would attend reclamation of the major derelict area in which a complex mixture of land misuses is present (Fig 37, and plates on pp 85 and 157), and here the technical difficulties are also considerable. The development of derelict land at Wincham–Marston was originally attributable to brine-pumping subsidence and the collapse of rock salt mines. By the 1870s (Fig 37) brine was also being pumped from flooded mines, and this led to particularly catastrophic damage and the formation of extensive flashes by the 1890s. The parallel decline of the salt industry within the locality and the rise of the chemical industry on its western subsidence-free margins added to the range of dereliction, for from the 1880s onwards chemical waste was dumped on the subsidence-damaged land. By the 1970s the greater part of the flashes had been converted into lime beds, and in addition there were the derelict sites of saltworks at several points along the line of the Trent and Mersey canal. Although no formal reclamation proposals exist for the whole of this area, it seems likely that the land will, in the long term, progressively be added to the Marbury Country Park, which is to be established in the derelict grounds of Marbury Hall immediately to the north (Collins 1972).

Derelict land and other planning problems in the North West

Dereliction is not a land-use problem to be seen in isolation, and reclamation schemes must be seen in the context of other planning decisions. There are few instances in the North West of major planning decisions that have been influenced by the prospect of reclaiming derelict land, although in this respect the region is little different from any other. The siting of Skelmers-

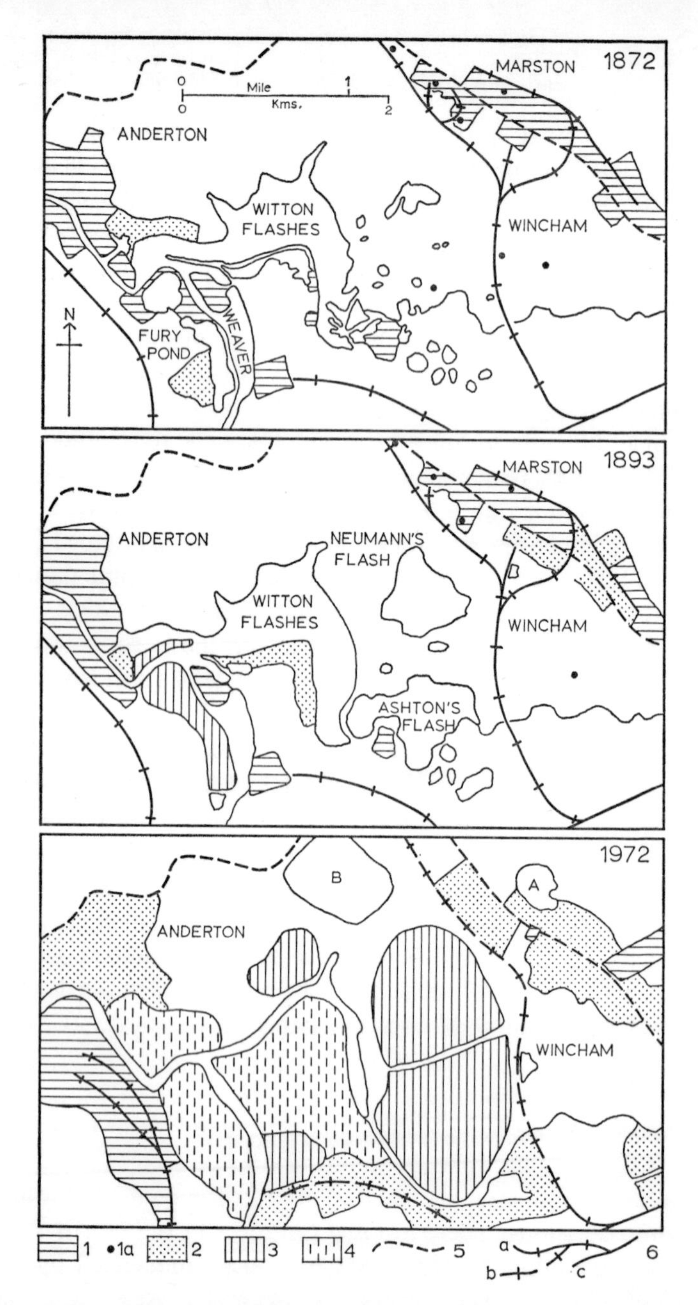

Fig 37 The growth of derelict land at Wincham-Marston, 1872–1972: 1, saltworks (and in 1972 only, chemical works); 1a, rock-salt mine; 2, derelict industrial land; 3, chemical waste; 4, vegetated chemical waste; 5, canal; 6a, mineral railway; 6b, abandoned mineral railway; A, site of Adelaide Mine; B, unused lime bed. Data for 1872 and 1893 from Ordnance Survey 1:2,500 plans; for 1972 from air photographs and field survey. See also Fig 2 and plate, p 157

dale New Town provided an opportunity to rehabilitate a moribund coal-mining area rather than build on first-class agricultural land. On the other hand the scale of reclamation involved was modest in comparison with that envisaged in a planning exercise outlining the development of new settlements on the worked-out exposed coalfield in south Lancashire (Reynolds 1966). The Winsford town expansion scheme provides an instance of an accelerated pace of reclamation following a major planning decision rather than being a factor in the original choice of location.

The most important general planning policy to have emerged from the conjunction of derelict land and other needs is related to the regional provision of recreational amenities. Of the schemes in being, the Wirral Country Park and the Rochdale Canal Park are both related to transport dereliction (see p 250), but partly completed projects and proposals such as those at Garswood Hall, Marbury Hall and Leigh, noted above, and the ambitous long-term scheme for the Croal-Irwell valley (Coates 1967), are based on areas of more complex derelict land. Schemes of this kind have the virtue that they simultaneously reclaim land and add to the stock of recreational facilities in densely populated areas poorly endowed with such amenities, while providing functional greenbelts within the major metropolitan areas. If the reorganisation of local government were to create a joint land reclamation agency, it could well achieve for the Liverpool–Manchester axis what the SVR has done for amenity planning in the Ruhr since the 1920s.

Dereliction forms part of a more general regional pattern of socio-economic difficulty. It is not possible to isolate the adverse influences of dereliction on the region's ability to attract new jobs or to hold its younger, more mobile inhabitants, but it is possible to compare the distribution of dereliction with other indices of socio-economic deprivation. In Fig 38 the areas with severest dereliction are plotted against those with poor smoke-control records and high pollution risks, and those with indifferent socio-economic health (Smith 1968). Only one small

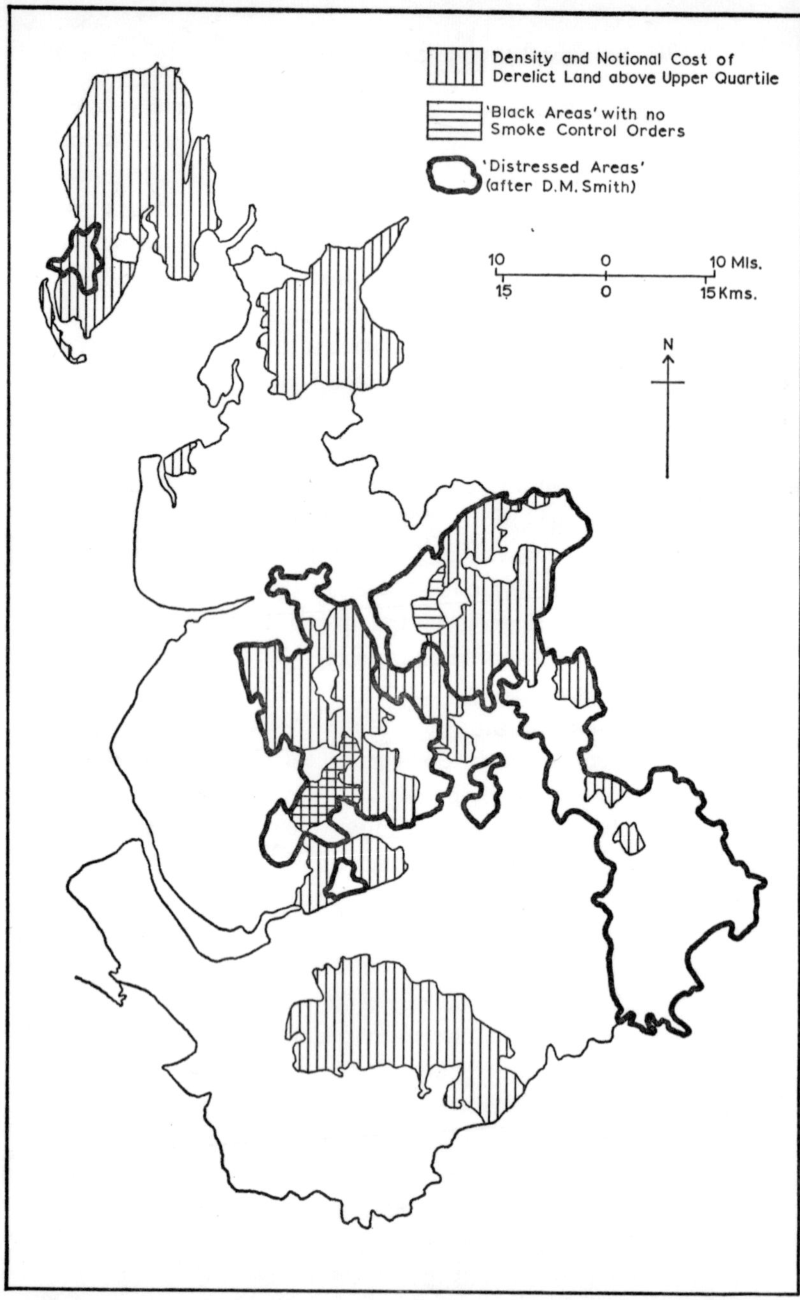

Fig 38 Problem areas in North West England. Data on derelict land derived from returns to the Department of the Environment, and on smoke control from the NW Economic Planning Council's report. D. M. Smith's map of 'distressed areas' appeared in *Regional Studies*, vol 2 (1968), 191. NB: This map differs in detail from Fig 33

area is badly placed on all counts, and not surprisingly it is Ashton–North Makerfield. Other areas with severe dereliction and poor socio-economic health include parts of the south Lancashire coalfield and Rossendale. But much of Furness and the whole of mid-Cheshire, both of them areas with severe dereliction, did not rate as 'distressed areas' in Smith's analysis. Conversely several localities in the Pennines, north east Lancashire and south-central Lancashire were 'distressed areas' but contained relatively little derelict land. An earlier survey of housing standards, confined to part of the North West (Burnett and Scott 1962), revealed that much of Rossendale and south Lancashire also possessed a poor housing stock. In spite of the progress of reclamation since the mid-1960s, the contention that south-central Lancashire's problems of environmental rehabilitation would take much longer to solve than its purely economic difficulties is still valid (Smith 1969). The census of 1971 supports this assertion, for parts of North Makerfield record poor levels of basic household amenities. Thus Ince in Makerfield UD and Aspull UD returned only 61 per cent and 63 per cent respectively of households with three basic amenities (exclusive use of hot water, fixed bath and indoor flush toilet); only Salford CB (58 per cent) recorded a lower proportion in Lancashire, where the county average was 78 per cent.

DERELICT LAND AND RECLAMATION IN THE WEST MIDLANDS

The West Midlands economic planning region contained 5,194ha of derelict land in 1969, at a density of 4ha per 1,000. However, 64 per cent of this derelict land was located within the geographical county of Staffordshire, at a density of 11ha per 1,000, and within Staffordshire 70 per cent of the dereliction was situated in the coalfield urban areas of North and South Staffordshire – the Potteries and the Black Country respectively. Both localities contained high densities of derelict land, although their rateable wealth reduced the levels of notional costs in the

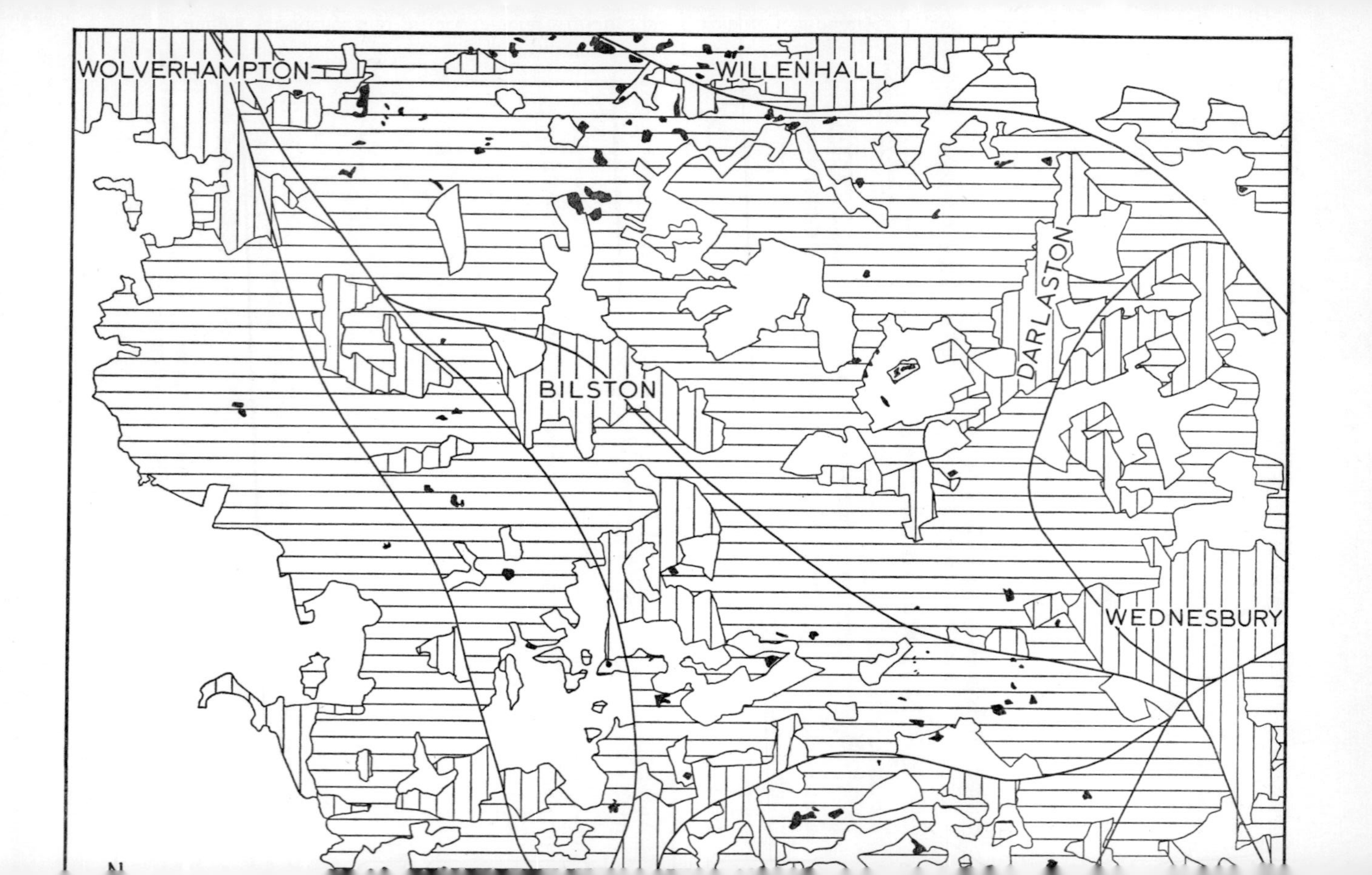

WOLVERHAMPTON
WILLENHALL
BILSTON
DARLASTON
WEDNESBURY

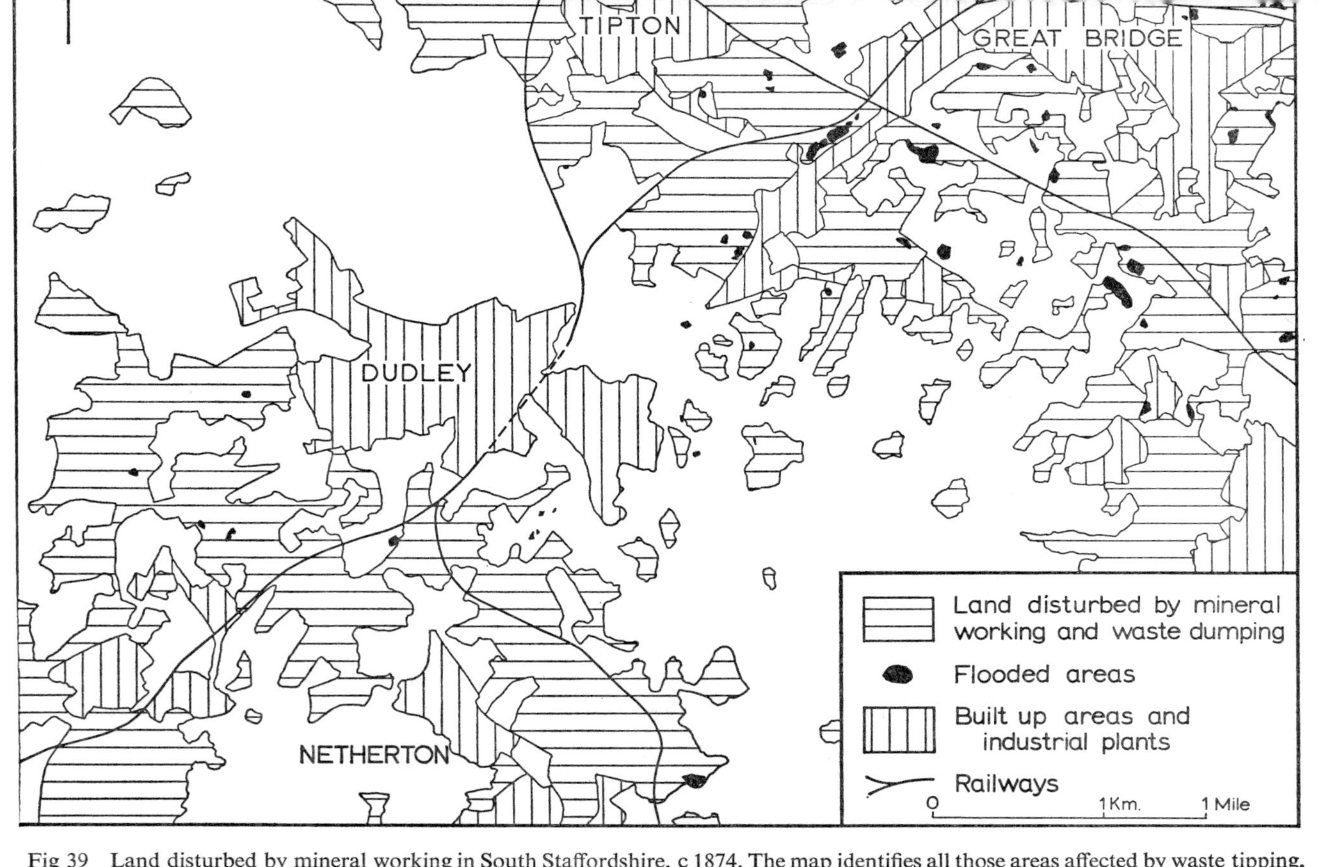

Fig 39 Land disturbed by mineral working in South Staffordshire, c 1874. The map identifies all those areas affected by waste tipping, mining subsidence and surface mineral working. Data from contemporary 1:10,560 Ordnance Survey maps

county boroughs (Fig 40). High densities of derelict land were also recorded in the Cannock Chase coalfield, where the forms of dereliction and the reclamation carried out resemble those in Lancashire (Staffordshire CC 1969), and in Cheadle RD, which contained abandoned coal pits, derelict copper smelting works, and disused pits and quarries. In this account it

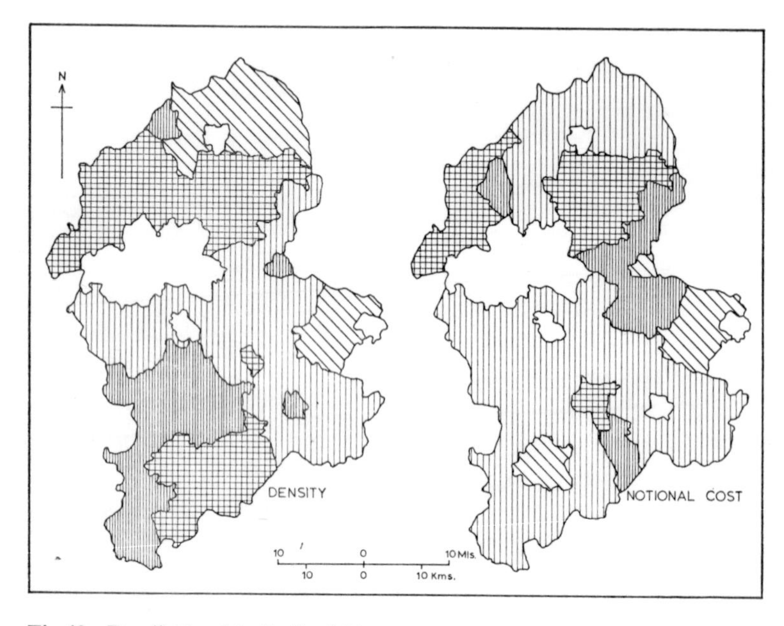

Fig 40 Derelict land in Staffordshire, 1969, showing density and notional cost of dereliction by local authority areas (see note below Fig 33)

is intended to deal solely with the Black Country and the Potteries, for in different ways these localities illustrate the development of derelict land and its reclamation in urban areas on a scale not repeated elsewhere in England.

Derelict land in the Black Country

The Black Country gained its reputation as an area of industrial pollution and dereliction during the first half of the nineteenth century, and by the 1870s much of the coalfield was disfigured by dumps of slag and spoil, hacked about by quarrying and clay digging, and pitted with subsidence hollows which, when flooded, were called 'swags' (Fig 39). The principal coal seam was the Thick Coal (9–11m thick) and this, together with other seams up to 3m thick, was worked at about 500 small collieries at mid-century. Coal was commonly worked by pillar and stall extraction; the supporting pillars were then mined, and in some instances a third phase of mining occurred when the consolidated debris was reworked. This pattern of mining produced large areas covered by low hummocks of spoil separated by subsidence hollows. In addition dereliction was caused by tipping blast furnace slag and chemical waste, the quarrying of limestone, sand, gravel, roadstone and brick clays, and the abandonment of industrial plants. The collapse of the traditional mining and heavy industries in the late 1880s contributed greatly to the spread of dereliction, which was estimated to cover about 5,670ha by 1903.

During the inter-war years piecemeal reclamation took place, much of it for housing and industrial development (Beaver 1945). The low amplitude 'hills and hollows' were relatively easy to restore, but even so, about 3,754ha of derelict land were recorded in 1945 (Beaver 1946a). Two-thirds of it was classed as 'chronic dereliction' and about one-quarter of the total was particularly intractable, including deep quarries and clay pits. Reclamation costs in general were estimated at £500–£865 per hectare, although this appears to have related to little more than levelling preparatory to further development.

The 1945 survey revealed the reclamation potential of much of the derelict land in the Black Country, particularly to meet the immediate post-war demand for housing sites. The amount of derelict land recorded in the Black Country fell from 3,933ha in 1946 to 1,554ha in 1965, a net decrease of 60 per cent (Table 27). The greater part of the reclaimed land was restored for housing (64 per cent), with industry accounting for a further 19 per cent. Since 1965 reclamation schemes have continued to erode the remaining dereliction, and of the once extensive 'hills

TABLE 27 Derelict land in the Black Country, 1946–65

	1946	*1954*	*1965*	*Net decrease 1946–65*
Dudley CB	332	123	121	211
Smethwick CB	7	5	4	3
Walsall CB	405	156	45	360
West Bromwich CB	401	201	212	189
Wolverhampton CB	176	63	16	160
Bilston MB	223	112	49	174
Halesowen MB	44	28	39	5
Oldbury MB	70	51	69	1
Rowley Regis MB	292	229	106	186
Stourbridge MB	30	15	5	25
Tipton MB	194	128	48	146
Wednesbury MB	211	162	67	144
Amblecote UD	26	} 399	23	3
Brierley Hill UD	424	}	225	199
Coseley UD	429	292	140	289
Darlaston UD	210	110	96	114
Sedgley UD	124	112	134	+10
Wednesfield UD	87	81	52	35
Willenhall UD	248	204	103	145
	3,933	2,471	1,554	2,379

Data from Oxenham (1966) for 1946 and from official returns to the Ministry of Housing and Local Government

and hollows' type only small pockets remain. Changes in local government boundaries in 1966 make it impossible to compare recent data with that for earlier years, but within the new area of the Black Country dereliction has continued to diminish (Table 28). The pattern of reclamation between 1947 and 1971 is shown on Fig 41, and the importance of restoration for residential and industrial uses can be seen. By 1971 the total amount of derelict land had fallen to 1,100ha, with major concentrations south-west of Dudley CB and north-west of Walsall CB.

The complex intermingling of derelict land and sites reclaimed

TABLE 28 Derelict land in the Black Country, 1966–71

Local authority area	*1966*	*1969*	*1971*	*Restored 1966–71*
Dudley CB	536	443	403	115
Walsall CB	312*	301	284	88
Warley CB	162	66	96	158
West Bromwich CB	384	249	233	193
Wolverhampton CB	91	101	84	63
TOTAL	1,485	1,160	1,100	617

* This is a revision of Walsall's original return on the basis of a major reappraisal carried out in 1968
Data from official returns.

at various dates is illustrated by the plate on p 176 and Fig 42. In the 1880s much of the locality was disfigured by disused collieries and abandoned clay pits and quarries, and in 1913 there was still extensive dereliction of this kind. During the inter-war period the new arterial road linking Birmingham with Wolverhampton was driven through the area, and much of the housing to the south of it was built on the sites of abandoned collieries. Similarly, the factories to the south of Dudley Port were built on land occupied by low spoil heaps and swags in 1921. After 1945 the pace of reclamation was maintained, and land was restored for industry, housing and educational uses: for example, Tividale School and its playing fields occupy the site of Hange Colliery, which had lain derelict for over 50 years. In 1971 little derelict land remained, but that which did included particularly intractable forms, such as the disused quarries at Derby's Hill. This typifies the Black Country generally, where the hard core of severe dereliction recognised in 1945 forms a high proportion of the surviving derelict land.

Derelict land in the North Staffordshire coalfield

The growth of derelict land in North Staffordshire is related partly to the activities of the various branches of ceramics manufacture and partly to other forms of extractive industry.

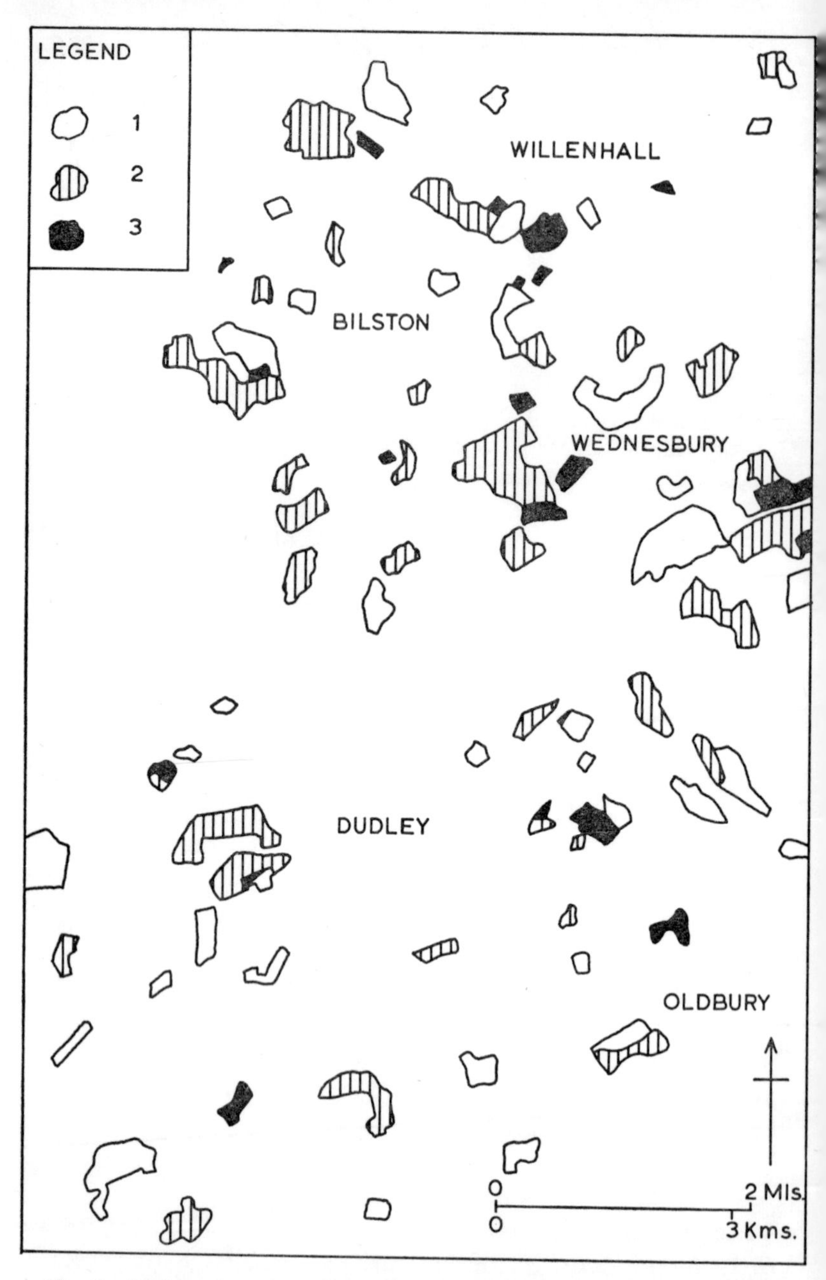

Fig 41 Reclamation of derelict land in South Staffordshire, 1947–71: 1, industry; 2, housing; 3, open spaces and recreation. Data from air photographs and topographical maps, checked for accuracy against a map produced by the West Midlands Regional Office, Department of the Environment, in 1972

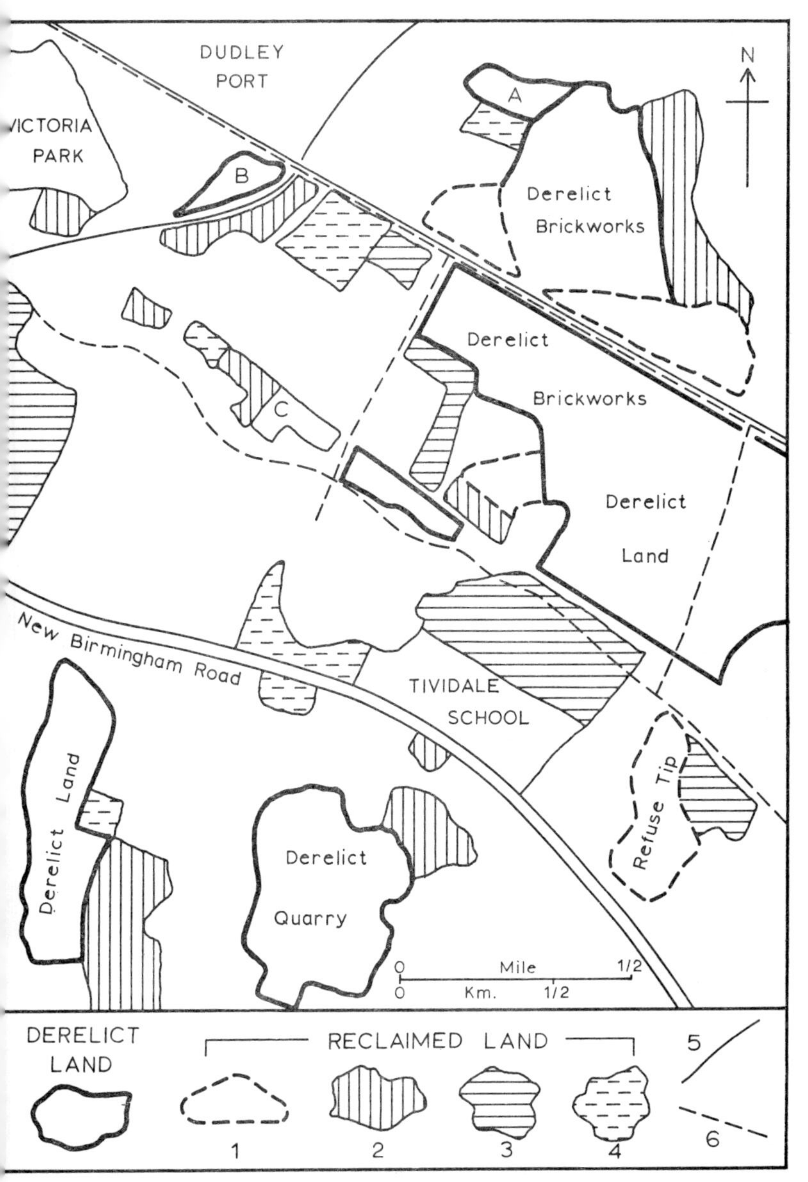

Fig 42 Derelict land and reclaimed sites in part of the Black Country, Staffordshire, October 1971. Land reclaimed since 1945: 1, work proceeding 1971; 2, housing; 3, industry; 4, open space and recreation; 5, railways; 6, canals. A, allotments; B, railway land; C, car park. (Victoria Park is an example of late-nineteenth century reclamation.) Based on an interpretation of plate, p 176; see also text, p 285

Since 1950 most of the dereliction caused by the ceramics industry has come from the abandonment of marl holes at tileries, and these deep excavations have proved to be particularly difficult to reclaim. In some instances, however, marl holes have been used as dumping grounds for shraff, the solid waste produced by potteries which commonly used to be tipped on the surface to create yet another form of dereliction. Iron mining and smelting were formerly causes of dereliction, but the slag heaps at abandoned ironworks have, as in the Black Country, found a ready market as road metal, and the slag produced at the surviving iron and steel works is immediately converted into tarred roadstone. Coal mining has been a more persistent source of dereliction, through the formation of spoil heaps (see plate, p 49) and the creation of flooded areas by subsidence. Although less damage from both sources is now likely, owing to the adoption of back stowage, there is a considerable backlog from earlier mining practices. Finally disused railways and mineral lines constitute an important element in the derelict scene.

Several accounts of the growth of derelict land in North Staffordshire have emphasised its intractability, particularly in comparison with that in the Black Country (Beaver 1946b, Moisley 1951), and this helps to explain the levels of dereliction revealed in Table 29. Nevertheless 650ha of derelict land were reclaimed in the area during 1954–67, about half of it for industrial and residential development. Valuable though this was, it represented a piecemeal attack on the problem, and in 1967 Stoke on Trent CB began to design a long-term reclamation programme in order to restore all the city's derelict land by the 1980s. Reference has been made to part of this programme on p 249, and it has also been the subject of several technical papers (Nicholson and Welmacott 1971, Plant *et al* 1971). Much of the initial expenditure of £770,000 to 1972 was to acquire and develop derelict land for amenity purposes, including Hanley Forest Park and Westport Lake. The scale of the programme gained great impetus from the inclusion of North Staffordshire in the North Midlands (sic) derelict land reclamation area. In contrast to the Black Country, North Staffordshire could not mount a large-scale long-term reclamation programme before state aid was made more widely available in 1966, and at a more

TABLE 29 Derelict land in the North Staffordshire coalfield, 1945–69

Local authority area	*Derelict land (ha)* 1945–46	*1954*	*1963*	*1969*	*ha per 000 (1969)*
Stoke on Trent CB	932	661	650	667	77·7
Newcastle under Lyme MB	443	432	369	304	84·7
Biddulph UD	30	30	19	17	6·3
Kidsgrove UD	223	176	134	120	72·8
Newcastle under Lyme RD	155	137	97	118	7·2
TOTAL	1,783	1,436	1,269	1,226	37·4

Data for 1945–6 from *North Staffordshire Plan* (1946), for 1963 from S. H. Beaver (1964), and for 1954 and 1969 from official returns

generous rate in 1970. The maintenance of state financial aid is clearly of the greatest importance if the ambitious plans of cities like Stoke on Trent to raise their environmental standards are to reach fruition.

References to this chapter begin on p 318.

CHAPTER TEN

Derelict Land: A Problem Capable of Solution?

RECOGNITION OF THE PROBLEM

THROUGHOUT this book dereliction has been regarded as a land-use problem, and present-day concern about environmental issues widely accepts this assertion. But it was not always so, and it is pertinent to ask when and why derelict land became a major public issue. The question is not easily answered, but in a British context a pointer is provided by the Parliamentary record. Parliamentary references to dereliction and related topics during the period 1929–71 have been tabulated (Table 30) on the basis of entries in the annual index to *Hansard*. The references relate to debates and answers to questions, the majority of the latter put to the responsible minister by members of Parliament on behalf of their constituencies. Before 1961 derelict land was not a common Parliamentary concern, whereas coal-mining subsidence and colliery spoil heaps were, although very little of this concern related to dereliction *per se*. Debates about subsidence related to the tangled problems of compensation rather than prevention of damage, and interest in colliery spoil heaps before 1939 largely centred on problems of atmospheric pollution; the revival of concern in the late 1960s related to tip safety in the aftermath of the Aberfan disaster. Thus, insofar as the Parliamentary record reflects public interest,

TABLE 30 Parliamentary references to derelict land

	References to:			
Parliamentary session	*Mining subsidence*	*Colliery spoil heaps*	*Derelict land*	*Related topics*
1929/30–1939/40	24	26	3	2
1944/45–1950/51	53(a)	4	1(c)	4
1951/52–1960/61	95(b)	25	6	16(d)
1961/62–1970/71	17	25	101	8(e)
TOTAL	189	80	111	30

(a) Largely related to Coal Mining Subsidence Bill of 1950
(b) Largely related to Coal Mining Subsidence Bill of 1957
(c) Adjournment debate on derelict land in the Black Country (*Hansard*, vol 423, columns 2361–8)
(d) Largely related to Mines and Quarries Bill of 1954
(e) Debates and questions on Aberfan disaster of 1966

References to opencast coal mining have been excluded, as the indexes do not consistently distinguish between entries relating to opencast mining and those relating to production, sales and quality of coal

Sources of data: Sessional Indexes to *Hansard*

derelict land did not become a problem until the 1960s and even then it did not feature prominently in debate.

A second guide to official concern is provided by Parliamentary enquiries and reports. Many of the enquiries into the state of the coal trade and conditions of work in the mines published before 1930 referred *en passant* to the causes of dereliction, but in the context of their influence on production rather than as problems in their own right. Similarly a major enquiry into coal mining techniques (1945) contained only a very brief report on methods of waste disposal, which was mainly concerned with the most expeditious and least expensive forms of surface tipping. Even where legislation was devised to tackle the causes of dereliction, it was hardly ever aimed at dereliction as a land-use problem of any consequence. This was particularly true of the Acts relating to coal-mining subsidence of 1950 and 1957, which made subsidence a respectable adjunct to mining. The outstanding exception was the creation of the Ironstone Restoration Fund in 1951, after exhaustive Parliamentary enquiries.

Before the 1960s there were few legislative aids to the reclamation of derelict land and some forms of active discouragement, exemplified by the attitude of the Commissioners for the Special Areas (p 213) and the operation of the 1944 Town and Country Planning Act (p 214). However, the period since 1960 has seen the enactment of more liberal measures, and current legislation greatly favours the restoration of derelict land with state financial aid (Table 31). The Hunt Report (1969) was particularly influential in suggesting that dereliction was 'one of the outstanding examples of the way in which an unfavourable environment can depress economic opportunity'. Although the report's recommendation that a national land reclamation agency should be established was not adopted, its view that grants-in-aid should be more widely available was accepted.

The idea that reclamation of derelict land should be encouraged as part of a policy of regional rehabilitation had been considered in relation to Central Scotland and North East England as early as 1963, but it was not until 1970 that the policy was more widely implemented. In 1972 a further departure from traditional practice occurred when the Special Environmental Assistance Schemes (and the equivalent Enterprise Ulster projects) were devised to reclaim derelict land and simultaneously provide jobs for the unemployed in the assisted areas of Great Britain and Northern Ireland respectively. Two other enquiries commissioned by Parliament in the early 1970s were also concerned to some extent with dereliction. That into environmental pollution (Ashby Committee 1970) regarded

TABLE 31 Legislation relating to reclamation of derelict land

Enactment	*Purposes of grant*	*Amount of grant per cent*	*Where available*
National Parks & Access to the Countryside Act, 1949 (together with Local Authorities [Land] Act, 1963)	Acquision and reclamation of derelict land required for the improvement of amenity	75	National Parks and Areas of Outstanding Natural Beauty
Industrial Development Act, 1966	Acquisition and reclamation of derelict land whose improvement will have a direct or indirect effect contributing towards the development of industry. (Subject to Department of Trade & Industry agreement)	85	Development Areas
Local Employment Acts, 1960 & 1970	As above	75	Intermediate Areas and Derelict Land Clearance Areas
Local Government Act, 1966	Acquisition and reclamation of derelict land for improvement of amenities	50	Universal

derelict land as a marginal issue, but that into mineral working (Stevens Committee, 1972) is likely to be of far greater relevance, in that it will almost certainly make recommendations on the restoration of mineral workings as a means of preventing further dereliction (see p 302).

Parliament is, however, not the sole mirror of public concern, and it seems likely that its increased attention to the problems posed by dereliction was a response to external pressures. While it is true that the Ministry of Town and Country Planning and its successor, the Ministry of Housing and Local Government (MHLG), had at times attempted to stimulate interest in derelict land, notably in the publication of *Derelict Land and Its Reclamation* (1956), no major impetus had been achieved before 1960. During the 1960s three publications in particular helped to arouse greater interest in the problem of dereliction than had been achieved by earlier technical memoranda and expert reports. *New Life for Dead Lands* (MHLG 1963), *Derelict Land* (Civic Trust 1964) and *The Countryside in 1970* (1967) all reached a wide audience, and must be considered important manifestations both of growing public concern and of increasing public pressure on the legislature that was reflected in various enactments during 1966–70.

ATTITUDES TO DERELICT LAND

Although it is evident that public awareness of the problems of dereliction was greater in the 1970s than ever before, as evidenced by the existence of extremely favourable governmental assistance towards the costs of reclamation schemes, it is difficult to know whether concern at governmental and local authority level was accompanied by greater individual awareness. Although no nationwide survey exists to support the contention, it seems likely that dereliction would rank low on most lists of environmental problems, which included items such as housing quality, atmospheric and water pollution, litter and noise. It is possible that it would not rank any higher in areas of dereliction, for it is claimed that the inhabitants of such localities display a 'derelict land mentality' which causes them to accept their surroundings as an inevitable accompaniment to mineral working.

R. A. Briggs (1970) has suggested that apathy can only be overcome by carrying out schemes of reclamation in order to demonstrate their practicability, and cites the restoration of Brancepeth tip, Willington, as a project which gave 'the townspeople a new self respect'; the same has been claimed of the Hanley Forest Park. Here comments made by the general public on further phases of the project revealed an interesting attitude to the colliery spoil heaps, which were regarded as local landmarks and, therefore, worthy of reshaping, rather than complete removal from the scene. It has even been suggested that the large-scale reclamation of derelict sites endangers the preservation of industrial monuments (Taylor 1973), and several recent schemes of amenity provisions have incorporated restored industrial sites, including the Llechwedd slate mine, Blaenau Ffestiniog and the Coalbrookdale district in Shropshire.

In a report on the Croal-Irwell valley (Coates 1967) it was observed that 'the unsightliness of the valley, universally accepted, was not subject to rigid classification, and could not be compared by the marshalling of quasi-objective data with any other depressed area . . . unsightliness and decay was in the eye of the beholder'. While agreeing that this must be so, it would be useful to know whether there is a consensus of opinion as to the varying degrees of acceptability attached to different types of derelict landscape.

Studies of landscape evaluation and of the perception of environmental problems do not as yet permit more than a simple analysis of attitudes, and the literature largely relates to pilot surveys, many of them based on the responses of students, who do not necessarily provide a representative opinion (O'Riordan 1971). A survey carried out during 1972 by the writer shares this defect by being based on responses from 380 students, who were asked to rank ten photographs of derelict landscapes on a seven-point scale of visual quality. In addition respondents were ask to name the most attractive and the least attractive of the landscapes, and to state what, if anything, they would regard

TABLE 32 Landscape evaluation

Photograph	*Average score**	*Percentage of respondents indicating landscape* (*a*) *most acceptable*	(*b*) *least acceptable*
1 Abandoned railway cutting	4·5	0·0	8·0
2 Disused lead mines, Grassington	4·0	3·0	2·5
3 Wet gravel pit (partly landscaped)	3·0	58·0	0·0
4 E. Midlands ironstone quarries	5·0	1·0	8·0
5 Flooded gravel pit (not landscaped)	5·5	0·0	35·5
6 China clay pit	4·0	6·5	4·0
7 Colliery spoil heaps	5·0	0·0	25·0
8 Blaenau Ffestiniog slate tips	4·5	3·0	5·0
9 Abandoned copper smelter	5·0	3·0	10·5
10 Abandoned Cornish engine-house	3·5	25·5	1·5

* On a seven-point scale, 1 represents superb landscape, 3 pleasant, 5 unpleasant and 7 hideous

as worse than the various scenes of dereliction. This last question is, by its open-endedness, likely to produce answers which defy simple classification, but it is interesting to note that roughly 60 per cent of the respondents did identify worse landscapes, mainly composed of active heavy industrial plants or features such as scrap-metal yards. Perhaps the proof that unsightliness is in the eye of the beholder came from the respondent resident in Merthyr Tydfil but studying in London, who considered 'the view of Hounslow High Street' to represent the least idyllic scene imaginable.

The results of the survey are summarised in Table 32, and five points may be made about them. Firstly, even the most acceptable of the landscapes rated an average score equated with 'pleasant' on the seven-point scale, and the remainder were rated 'neutral' to 'ugly'. Second, the setting of an abandoned mineral working greatly affected responses to its landscape quality: of the two photographs of abandoned wet gravel pits, that with a background of trees, resembling a lake in an ornamental park, was considered to be the most acceptable view by 58 per cent of the respondents. Conversely that with a background of anonymous sheds and a sky criss-crossed by power lines was rated least acceptable by 35·5 per cent of the respondents. Thirdly, the most dramatic forms of mineral working (eg slate quarries and china clay pits) evoked mixed reactions, even though the landscapes rated badly on the seven-point scale. Thus more respondents found the china clay pits the most acceptable scene than considered them to be the least pleasing, and the converse was true, by a similarly narrow margin, of the slate quarries. Fourthly, there was far less unanimity about the least acceptable scene but in this context it is interesting to note that the photograph of colliery waste heaps ranked a close second in the range of unattractive scenes. Finally, on the evidence of this survey those who categorised abandoned Cornish engine-houses as an attractive form of dereliction did so with good reason (or, possibly, helped to condition the respondents' answers!). Although the results of this survey are very tentative, they do at least suggest that attitudes to derelict land are not clear-cut, and that the observation made in the Croal-Irwell valley report has a more general validity.

THE REASONS FOR RECLAMATION

The main purpose of recognising derelict land as a problem is to consider means of restoring it to beneficial use. The surveys of dereliction carried out in England during 1964–71 provided a basis for identifying probable reclamation projects, and the new survey of derelict and despoiled land introduced in 1973–4 will identify potential as well as existing problem areas. But why should it be thought important to reclaim derelict land, particularly that which has lain idle for 40 years or more without great public concern being evinced about its removal?

In an economic sense reclamation of derelict land might appear justified in that it restores abandoned or under-used sites to profitable uses. But it is doubtful whether many of the reclamation schemes carried out in recent years for agricultural and general amenity purposes will ever show a satisfactory yield on capital expenditure (Wibberley 1960), and the net addition to agricultural production could have been achieved at less cost by up-grading existing farm land. There may be instances in which the addition of reclaimed land to an individual farm holding may make it a more economic enterprise, but in general reclamation for agriculture, including that consciously carried out as part of the mineral-working process, as at opencast coal mines, is uneconomic. Reclamation for other purposes which will yield little or no immediate revenue, notably various forms of recreational use, may be justified on the grounds that it provides an alternative to taking agricultural land out of production. But the argument that land lost to farming for urban uses causes irreparable harm to agriculture is difficult to sustain, for it is often possible to make good the deficiency by up-grading other farm land. Thus 'saving' farm land by providing playing fields on a reclaimed derelict site may prove to be an expensive and illusory form of protecting agricultural production.

Reclamation for industrial and residential uses may be economically more worthwhile if only because the after-value of the reclaimed site sometimes exceeds the gross cost of reclamation. But this is not universally so, and it would be as uneconomic to build houses on an expensively reclaimed site in the belief that this protected agricultural production as it would to convert the same derelict land to playing fields or to pasture. The economic argument for reclaiming derelict land for urban uses is most likely to hold within existing urban areas where there may be little developable land, and the need to provide space for recreation and amenity can often be met most cheaply by reclaiming derelict sites, even though the cost appears to be staggeringly high (see p 231–2). The additional cost of building houses on derelict land in the Black Country, estimated at £850 per ha in 1946, was considered to be worthwhile because no acceptable alternative existed to providing houses within this area of great industrial importance. Similarly the reclamation of derelict land for industrial development may be economically justifiable if the alternatives are either to curtail the expansion of factory production or to locate part of the plant on a more distant site.

Many protagonists of the derelict-land reclamation programme would not seek to justify it in readily quantifiable economic terms. Several of the earliest schemes of reclamation for industrial building, in South Wales and Tyneside, were socially motivated, and before the 1960s much of the land reclaimed for industrial development in such localities formed part of a social rather than a purely economic policy. In recent years much more attention has centred on reclamation which may, by improving the physical environment, attract new forms of investment and, in doing so, hold those young inhabitants who show the greatest propensity to migrate. Both the Lancashire and West Riding of Yorkshire county planning offices have justified parts of their reclamation programmes on these grounds, among others. Whether would-be industrialists were deterred from establishing factories in Yorkshire because the view from the motorway was punctuated by the prospect of colliery spoil tips was a debatable point that was not, however, readily accepted by the central government. An ambitious project designed to landscape the motorway corridor was not granted the highest rate of aid because no positive relation could be demonstrated between the restoration of derelict land and the provision of new jobs in the locality (Warren 1972). Similarly the assumption that young migrants leave derelict areas because they *are* derelict rather than lacking in other amenities merits closer questioning than it has generally been given. However, there is some evidence to suggest that reclamation schemes engender greater civic pride, and this in itself may bring long-term socio-economic benefits to a locality.

There can be little doubt that one of the greatest stimuli to reclamation has come from granting state aid to amenity schemes, and these may ultimately produce more tangible economic benefits. The general concern to provide better facilities for leisure pursuits, particularly within easy access of major urban centres has given a spur to reclamation projects which would not normally have come from market demands for industrial and residential sites in many areas of dereliction.

An assessment of the total expense of removing dereliction from the face of Great Britain is difficult, and it will never be possible to calculate the social costs already incurred by permitting derelict land to exist in such large quantities. The central government is committed to an attempt to remove the backlog of dereliction by the early 1980s, but many local authorities have taken a more cautious view in framing their programmes of reclamation. At a conservative estimate it would cost £126 million to remove the 39,292ha of derelict land recorded in England alone at the end of 1971, but if, as many surveys suggest, the true extent of dereliction is two to three times as great as the official total, the cost would need to be amended accordingly. As noted in previous chapters costs of reclamation vary appreciably and it may be that the most intractable, and by inference the most costly, schemes form a disproportionately large percentage of the land awaiting treatment.

A final problem in estimating the financial cost of reclamation centres on the extent to which private capital will fund the restoration of derelict land. The Cheshire county planning office estimated that at least 15 per cent of its derelict land would be restored by private enterprise, notably in developing former military establishments as sites for housing and industry, but it is doubtful whether this proportion could be extrapolated to cover all derelict land. On the evidence of the 1972 returns (Table 13) privately financed reclamation was least common in those localities with the most severe problems of dereliction, and it seems likely that public finance will continue to be necessary in order to maintain the progress of reclamation. Compared with other elements of public expenditure, an outlay of from £250 million to £380 million over 20 years or so on the clearance of all derelict land, official and unofficial, does not seem excessive. (At 1970 rates a sum of £250 million spread over 20 years would be equivalent to 0·5 per cent of public expenditure on housing and environmental services, excluding interest charges.) Whether such a reclamation programme is technically feasible in view of the pressures under which many planning offices work is open to question. Similarly opinions differ as to whether all derelict land merits restoration. More important is whether the reclamation of the derelict land justifying treatment in 1971 marks the end of the task, or is simply a phase in a continuing attempt to stem the growth of dereliction.

PREVENTION OF FURTHER DERELICTION

The attempt to reclaim all derelict land by the early 1980s presupposes that the rate at which new dereliction is created will fall dramatically during the next decade or so. The same hope may well have been entertained when the 1947 Town and Country Planning Act passed on to the statute book, for it provided the means both to control future mineral working and to include conditions of restoration in individual planning consents. There were, however, exclusions from planning control under the General Development Order, the main effect of which was to exempt existing workings and extensions to them. Even where planning conditions have been imposed under the 1947 Act and its successors, it has not always proved possible to enforce them for a variety of legal and technical reasons.

Many of the early planning conditions were imperfectly or imprecisely framed, but the growth of expertise in designing planning controls for mineral extraction has not always led to a successful conclusion. There are four main reasons for this.

Firstly, filling materials referred to in a planning consent may not be available when reclamation is to be carried out. This has been particularly true of schemes which were to have relied for fill on PFA, for which more profitable uses have been found, but it has also occurred where costs of transporting fill had been underestimated. Secondly, the geological conditions of workings may not accord with original expectations, particularly if variations in water table affect the mode of extraction or if the amount of waste remaining after completion of working differs from that covered by the restoration conditions. Thirdly, uncertainties of demand may cause workings to be 'temporarily closed', so that they appear to be derelict but may in fact be reopened when demand revives. Fourthly, mineral firms may reorganise their operations, and this also commonly leads to premature 'temporary closure', which prolongs the period of dereliction. Bankruptcy and insolvency have the same effect as amalgamations and other forms of reorganisation, with the added problem that the former owner will have no financial resources to fund even the simplest restoration scheme. These observations apply almost wholly to surface mineral working in Great Britain, with the exception of ironstone and coal, which, as nationalised industries covered by special statutory regulations, are normally dealt with through other planning provisions.

Clearly the prevention of future dereliction depends on ensuring that mineral working is more strictly controlled and that restoration is funded as a normal production cost rather than a state-financed afterthought. In 1972 the Department of the Environment and other government departments established a joint committee of enquiry (the Stevens Committee)

> . . . to examine the operation of the statutory provisions (except the provisions of the Opencast Coal Act 1958) under which planning control is exercisable over mineral exploration, over surface mineral working and installations, over the deposits on the surface of spoil or waste from mineral workings, and over the after-treatment of surface land worked for minerals; to consider whether the provisions require to be amended or supplemented; and to make recommendations.

The committee will probably report in 1974, but it seems likely to consider the four strategies listed below as alternative ways of reducing future dereliction. The comments which follow are based on the opinions of county planning officers with considerable experience of minerals planning:

1 a levy on mineral production in order to finance a central reclamation fund similar in principle to the Ironstone Restoration Fund;
2 a reclamation bond deposited by the operator to cover the anticipated cost of restoration;
3 fixed penalties for failure to comply with planning conditions, coupled with the withholding of any further planning permission from the operator;
4 comprehensive planning laws, whereby the absolute right to extract minerals is exchanged for an equally firm obligation to restore sites after completion of working.

Restoration levy

The success of the Ironstone Restoration Fund almost certainly reflects its confinement to a single industry under a handful of operators working within a relatively small area. If the same principle were to be more widely applied, it might prove difficult to devise a rate of levy which would be appropriate to particular sections of the mineral industry. Were a flat rate based on tonnage raised to be introduced for all mines and quarries, the operators of, say, dry gravel pits might find themselves financing the restoration of slate quarries. Even if the levy was varied according to type of working, the range of reclamation costs would mean that an operator's contribution to the fund could be devoted to reclaiming the land of a less efficient competitor. Few planning authorities favour the levy, but, significantly, one major group, forming the East Midlands regional development conference, has produced a reasoned report favouring its introduction (1972). The report envisages that the levy would be based on the rateable value of workings, which, among other things, is based on tonnage raised and, unlike the valuation of other property, is revised annually.

Rateable value is held to provide an acceptable measure of liability, but it is not an adequate predictor of reclamation costs. These vary greatly from site to site, even within the same industry over a small locality, and this variation could prove to be a major difficulty in introducing any form of levy.

Reclamation bond

The reclamation bond in the form of cash deposits or, more probably, sureties, is favoured by several local authorities (at least two British planning authorities already operate such a system), and it also figures prominently in recent US mineral legislation. The strength of this device is thought to lie in the fact that it can be tailor-made for each individual case and therefore avoids the possible inequalities of the flat-rate levy. Arguments against the bond centre on two points: firstly, the costs calculated at the onset of working may be affected by subsequent operational difficulties (see p 302), or they may be distorted by inflation; secondly, small concerns may not have the capital to finance an adequate bond, or income to secure insurance sureties.

Penalty system

This is incorporated in recent US legislation and in effect forms part of British planning law, in that failure to comply with restoration conditions after the serving of an enforcement order may lead to a fine. In fact such procedures are rarely resorted to in Great Britain because the offending firm is more likely to exploit loopholes in the law than to comply with the planning requirements under duress. Because of this no local authority favours the introduction of a penalty system on the American pattern, but several would welcome a strengthening of existing planning laws, which would have the same effect.

Comprehensive mineral laws

This alternative is suggested by the planning and mineral working laws which operate so successfully in the Cologne brown-coal field (p 179). The opencast executive of the NCB has carried out comprehensive extraction and restoration of a similar kind but on a smaller scale, notably in Northumberland (Bennett 1969). Although some find the prospect attractive, most planning authorities do not see similar legislation to that enforced in Germany finding favour in Great Britain (and it has to be remembered that German law took many years to evolve and has worked most successfully in the sparsely populated forested areas of the Ville).

The two methods of control that are most likely to be employed in the attempt to prevent the creation of further dereliction are, therefore, the restoration levy and the restoration bond, possibly as complementary modes of securing finance from different sections of the mineral industry. The precise form to be adopted depends on the solution of a complex range of technical and legal issues, assuming that the principle of restoration immediately after working has ceased is accepted (and the definition of the difference between completion and temporary stoppage of working in itself presents a fundamental legal difficulty). It is, however, widely agreed that in the interests both of land-use planning and of effective mineral working there should be a major revision of the law which would bring, as one of its benefits, the prevention of newly created dereliction (Thorburn 1973). There would still remain a substantial backlog of existing dereliction to be treated, including that which does not currently qualify for inclusion in the official returns but is nevertheless derelict land which would not be covered by any kind of fund envisaged in future legislation. Thus the problem of dereliction would not disappear overnight, but its continued growth would be prevented.

CONCLUSION

This book began by examining some of the problems of identifying derelict land and has ended on a similar note by investigating the difficulties of controlling mineral extraction within the framework of British planning law. But the greater part of this study has been concerned with the origins and development of derelict land and the recent progress made in restoring it to beneficial uses. There can be little doubt that the technical problems of reclaiming derelict land no longer present the major obstacles to progress. The initial civil engineering works can be successfully programmed on the basis of accumulated experience and there is a substantial fund of knowledge about drainage techniques, mode of soil formation and tolerance of plant species on various types of reclaimed land. Although less is known of satisfactory methods of after-management, this aspect of reclamation is also being tackled energetically, so that, once restored, land will not degenerate into near-dereliction within a few years of treatment.

The financial obstacles to reclamation also appear to have been more successfully overcome than seemed possible less than a decade ago, although it is argued that a higher rate of state aid would be welcome outside the Development Areas. The abandonment of narrow economic criteria in favour of a more liberal view of the socio-economic benefits which may stem from restoration was a particularly enlightened change of financial policy in the 1960s. There are, however, still obstacles to the complete eradication of derelict land. Public awareness and concern is perhaps least evident in those localities which might secure most benefit from a sustained programme of restoration. Even so, there are signs that the mounting of large-scale restoration projects has engendered civic pride among the inhabitants of localities long notorious for the visual squalor of their dereliction, indicating that the 'derelict-land mentality' is not a fixed state of mind. The most important problem awaiting solution is that of framing adequate legislation to ensure that restoration will automatically follow the completion of mineral working. It must be remembered that much of the least tractable derelict land is the product of recent economic development rather than a legacy of earlier primitive phases of exploitation. It would be unfortunate if the proposals to remove existing derelict land during the 1970s obscured the fact that action is even more urgently needed to prevent the creation of further dereliction. The findings of the Stevens Committee may well lead to legislation which would demonstrate the practicability of finally solving the problems of dereliction in Great Britain.

References to this chapter begin on p 320.

References and Bibliographies

SPECIFIC notes have not been included in the text, but references to books or readily identifiable chapters within them, and to papers in professional and trade journals, have been cited and are listed below for each chapter. In some instances additional references relating to the contents of individual chapters are also provided. No bibliography is provided for two reasons. Firstly, the chapter references themselves provide a considerable amount of bibliographical material, and secondly, there already exists a range of bibliographies, which are cited as additional references to Chapter 10. Material on derelict land and its restoration published by local authorities and other interested parties is commonly listed in *The Planner* (the journal of the Royal Town Planning Institute), and in the guides to current literature produced by the library of the Department of the Environment, to both of which reference should be made.

CHAPTER ONE

Beaver, S. H. *Derelict Land in the Black Country* (Ministry of Town and Country Planning, 1946)
—— 'Land reclamation after surface mineral workings', *Journal of the Town Planning Institute*, vol 41 (1955), 146–52
—— 'Changes in industrial land use, 1930–67', in *Land Use and Resources: Studies in Applied Geography* (1968), 101–10
—— 'An appraisal of the problem', *Proceedings of the Derelict Land Symposium* (1969), 3–9
—— 'The reclamation of industrial waste land for agriculture and other purposes', in *Problems of Applied Geography* (Warsaw, 1961), 97–112
Best, R. H., and Coppock, J. T. *The Changing Use of Land in Britain* (1962)
Best, R. H. 'The extent of urban growth and agricultural displacement in post-war Britain', *Urban Studies*, vol 5 (1968), 1–23 (a)
—— 'Competition for land between rural and urban uses', in *Land Use and Resources: Studies in Applied Geography* (1968), 89–100 (b)
Bush, P. W. 'Spoiled lands to the south-east of Leeds', *Proceedings of the Derelict Land Symposium* (1969), 21–8
Civic Trust. *Derelict Land* (1964)
—— *Reclamation of Derelict Land* (1970)
Collins, W. G., and Bush, P. W. 'The definition and classification of derelict land', *Journal of the Town Planning Institute*, vol 55 (1969), 111–15
Hoskins, W. G. *The Making of the English Landscape* (1955)
Jacobs, C. A. J. *Derelict Land in Denbighshire* (Ruthin, 1972)
North West Economic Planning Council. *Derelict Land in the North West* (Manchester, 1969)
Oxenham, J. R. 'The reclamation of derelict land', *Journal of the Institution of Municipal Engineers*, vol 75 (1948), 182–202
—— *Reclaiming Derelict Land* (1966)
Penning-Rowsell, E. C., and Hardy D. I. 'Landscape evaluation and planning policy: a comparative survey in the Wye Valley AONB', *Regional Studies*, vol 7 (1973), 153–60

CHAPTER TWO

For additional material see:

Beaver, S. H. 'Minerals and planning', *Geog J*, vol 104 (1944), 166–93

—— 'An appraisal of the problem', *Proceedings of the Derelict Land Symposium* (1969), 3–9

Most of the detailed references to specific industries are cited at the end of chapters in Part Two. The maps used in compiling Figs 11 and 12 are contemporary copies of the Ordnance Survey 1:63,360 series, revised at various dates within the periods 1888–1904 (but mainly 1890–1900) and 1946–57 (but mainly 1948–54) respectively. Fig 10 is based on an analysis of the David & Charles facsimiles of the 1:63,360 maps, which relate to the 1840s except for the sheets covering Northumberland and Durham, surveyed mainly during the 1850s

CHAPTER THREE

Beaver, S. H. *Northamptonshire*, part 58 of *The Land of Britain* (1943)

Coleman, A. 'Land reclamation at a Kentish colliery', *Transactions and Papers, Institute of British Geographers*, No 21 (1955), 117–35

Du Maurier, D. *Vanishing Cornwall* (1967)

Holmes, W. D. 'The Leicestershire and South Derbyshire Coalfield', *East Midland Geographer*, vol 2, pt 10 (1958), 16–26; pt 12 (1959), 9–17; pt 13 (1960), 16–21

Hoskins, W. G. *Leicestershire* (1957)

Ministry of Housing and Local Government (MHLG). *Derelict Land and Its Reclamation*, Technical Memorandum, No 7 (1956)

Priestley, J. B. *English Journey* (1934)

Robertson, I. M. L. 'Occupational structure and distribution of rural population in England and Wales', *Scottish Geographical Magazine*, vol 77 (1961), 165–79

Wallwork, K. L. 'Land-use problems and the evolution of industrial landscapes', *Geography*, vol 45 (1960), 263–75

—— 'Minerals and the problem of derelict land in the East Midlands', *East Midland Geographer*, vol 6 (1974), 17–28

CHAPTER FOUR

Blunden, J. R. 'The renaissance of the Cornish tin industry', *Geography*, vol 55 (1970), 331–3

Carr, J. P. 'The rise and fall of Peak District lead mining', in Whittow, J. B., and Wood, P. D. *Essays in Geography for Austin Miller* (1965), 207–24

Cole, M. M. 'The Witwatersrand conurbation: a watershed mining and industrial region', *Transactions and Papers, Institute of British Geographers*, 23 (1957), 249–65

Collins, J. F. N. *A Policy for the Control of Salt Extraction in Cheshire* (Chester, 1971)

Collins, W. G., and Bush, P. W. 'Aerial photography and spoiled lands in Yorkshire', *Journal of the Royal Town Planning Institute*, vol 57 (1971), 103–10

Collins, W. G., and James N. 'Spoiled land: an assessment, with aerial photographs, of its suitability for development in West Cornwall', *The Planner*, vol 59 (1973), 444–9

Denison, W. J. 'Devising and assessing reclamation schemes', *Proceedings of the Derelict Land Symposium* (1969), 17–20

Edwards, D. B., and Rees, G. V. *Derelict Land Report: Flintshire* (Mold, 1973)

Elkins, T. H. 'Liége and the problems of Southern Belgium', *Geography*, 41 (1956), 83–98

Goodridge, J. C. 'The tin-mining industry: a growth point for Cornwall', *Transactions and Papers, Institute of British Geographers*, No 38 (1966), 95–104

Grey, R. E., and Meyers, J. F. 'Mine subsidence and support methods in the Pittsburgh area', *Proceedings American Society of Civil Engineers, Soil Mechanics Division*, vol 96 (1970), 1,267–92

Harris, A. *Cumberland Iron: The Story of Hodbarrow Mine, 1855–1968* (Truro, 1971)

Hoyle, B. S. 'The production and refining of indigenous oil in Britain', *Geography*, vol 46 (1961), 315–21

Hull, E. 'On the recent remarkable subsidences of the ground in the salt districts of Cheshire', *Royal Dublin Society Sci Proc*, 3 (1883)

Institution of Civil Engineers. *Mining Subsidence* (1959)

Klimaszewski, M. 'The problems of the geomorphological and hydrographic map on the example of the Upper Silesian industrial district', in *Problems of Applied Geography* (Warsaw, 1961), 73–82

Legget, R. F. *Cities and Geology* (New York, 1973)

Meiklejohn, J. 'The Klondike Gold-fields', *Transactions of the Institution of Mining Engineers*, vol 19 (1900), 352–64

National Coal Board. *Mining Subsidence Engineer's Handbook* (1965)

Niddrie, D. L. 'Uranium from the Union of South Africa', *Geography*, vol 40 (1955), 193–4

Oxenham, J. R. *Reclaiming Derelict Land* (1966)

Pocock, D. C. D. 'Britain's post-war iron-ore industry', *Geography*, vol 51 (1966), 52–5

Poland, J. F., and Davis, G. M. 'Land subsidence due to withdrawal of fluids', in Detwyler, T. R. *Man's Impact on Environment* (New York, 1971), 370–82

Pounds, N. J. G. 'The spread of mining in the coal basin of Upper Silesia and Northern Moravia', *Annals Assocn Amer Geogrs*, vol 48 (1958), 149–63

Raistrick, A. *The West Riding* (1970)

Robinson, A. H. W., and Wallwork, K. L. *Map Studies with Related Field Excursions* (1970)

Siedlungsverband Ruhrkohlenbezirk (SVR). *Regionalplanung* (1960, Essen)

Statham, D. C. 'The development of the Yorkshire potash mining industry', *Town Planning Review*, vol 42 (1971), 361–76

Thomas, T. M. *The Mineral Wealth of Wales and its Exploitation* (1961)

Vogel, I. *Bottrop, Ein Bergbaustadt in der Emscherzone des Ruhrgebietes* (Remagen, 1959)

Wallwork, K. L. 'Subsidence in the Mid-Cheshire industrial area', *Geogr J*, vol 122 (1956), 40–53

—— 'The Mid-Cheshire salt industry', *Geography*, vol, 44 (1959), 171–86

—— 'Some problems of subsidence and land use in the Mid-Cheshire industrial area', *Geogr J*, vol 126 (1960), 191–9 (a)

—— 'Land-use problems and the evolution of industrial landscapes', *Geography*, vol 45 (1960), 263–75 (b)

—— 'Map interpretation and industrial location: the example of alkali manufacture in Lancastria', *Geography*, vol 52 (1967), 166–81

—— 'Salt and environmental planning: an historical perspective to a contemporary land-use problem', *Proceedings, Fourth Symposium on Salt* (Houston, 1973)

Ward, T. 'The rock-salt deposits of Northwich, Cheshire, and the result of their exploitation', *Trans Manch Geol Soc*, vol 25 (1898), 274–306 and 530–67

—— 'The subsidences in and around the town of Northwich in Cheshire', *Transactions of the Institution of Mining Engineers*, vol 19 (1900), 241–64

Wohlrab, R. 'Einwirkungen des Bergbaues auf Wasserhaushalt und Landwirtschaft unter besonderer Berücksichtigung der deutschen Verhältnisse', *Berichte zur Deutschen Landeskunde*, 39 (1967), 81–100

CHAPTER FIVE

Abercrombie, Sir L. P. *Greater London Plan 1944* (1945)

Barton, P. M. *A History of the Cornish China-clay Industry* (Truro, 1966)

Barton, D. B. *A History of Tin Mining and Smelting in Cornwall* (Truro, 1967)

Beaver, S. H. 'Minerals and planning', *Geogr J*, vol 104 (1944), 166–93

—— 'The reclamation of industrial waste land for agricultural and other purposes', in *Problems of Applied Geography* (Warsaw, 1961), 97–112

—— 'An appraisal of the problem', *Proceedings of the Derelict Land Symposium* (1969), 3–9

Bennett, P. M. 'Restoration and reclamation of land by the Opencast Executive of the NCB', *Proceedings of the Derelict Land Symposium* (1969), 38–41

Boden, P. K. 'The limestone quarrying industry of north Derbyshire', *Geogr J*, vol 129 (1963), 53–62

Coleman, A. 'Landscape and planning in relation to the cement industry of Thames-side', *Town Planning Review*, vol 25 (1954), 216–30

Coudé, A. 'La mise en valeur des tourbières et l'utilisation de la tourbe en République d'Irlande', *Annales de Géographie*, vol 87 (1973), 576–605

Cowley, G. *Bedfordshire Brickfield: A Planning Appraisal* (Bedford, 1967)

—— *Bedfordshire County Review: Minerals Aspect Report* (Bedford, 1972)

Crompton, C. T. 'The treatment of waste slate heaps', *Town Planning Review*, vol 38 (1967), 291–304

Cumberland, K. B., and Whitelaw, J. S. *The World's Landscapes: 5, New Zealand* (1970)

Cummings, L. P. 'The Bougainville copper industry construction phase', *Australian Geographer*, vol 12 (1972), 55–6

Dale, R. W. *Peat in Central Somerset: A Planning Study* (Taunton, 1967)

Davis, J. F., and Baker, R. F. 'The economic geography of lower Thames-side', in Clayton, K. M. (ed). *Guide to London Excursions* (1964), 64–8

Dwyer, D. J. 'The peat bogs of the Irish Republic, a problem in land use', *Geogr J*, vol 128 (1962), 184–93

Elkins, T. H. 'The brown-coal industry of Germany', *Geography*, vol 38 (1953), 18–29 (b)

—— 'The Cologne brown-coal field: a study of the law and the landscape', *Transactions and Papers, Institute of British Geographers*, No 19 (1953), 131–43 (a)

—— 'East Germany's changing brown-coal industry', *Geography*, vol 41 (1956), 192–5

Forrest, J. 'Otago during the gold rushes', in Waters, R. S. (ed). *Land and Society in New Zealand* (Wellington, 1965), 80–100

Förster, H. 'Das nordböhmische braunkohlenbecken', *Erdkunde*, vol 35 (1971), 278–91

Grimshaw, P. N. 'The U.K. Portland cement industry', *Geography*, vol 53 (1968), 81–4

Hall, W. G. (ed). *Man and the Mendips* (1971)

Jacobs, C. A. J. *Derelict Land in Denbighshire* (Ruthin, 1972)

—— *Quarrying and the Environment: Study of Bulk Mineral Extraction in Denbighshire* (Ruthin, 1973)

Johnson, J. H. 'Commercial use of peat in Northern Ireland', *Geogr J*, vol 125 (1959), 398–400

Kohn, C. F., and Specht, R. E. 'The mining of taconite, Lake Superior Mining District', *Geographical Review*, vol 48 (1958), 528–39

Orme, A. R. *The World's Landscapes: 4, Ireland* (1970)

Pocock, D. C. D. 'Iron and Steel at Corby', *East Midland Geographer*, vol 2, pt 15 (1961), 3–10

—— 'Stages in the development of the Frodingham Ironstone field', *Transactions and Papers, Institute of British Geographers*, No 35 (1964), 105–18

Robinson, A. H. W., and Wallwork, K. L. *Map Studies with Related Field Excursions* (1970)

Scott, P. 'The brown coal industry of Australia', *Geography*, vol 43 (1958), 212–15

Udall, S. A. *A Study of Strip and Surface Mining in Appalachia* (Washington, DC, 1966)

United States Department of the Interior (USDI). *Surface Mining and Our Environment* (Washington, DC, 1967)

Wallwork, K. L. 'Mining, quarrying and derelict land: an aspect of land use in Northern Ireland', *Irish Geography*, vol 6 (1973), 570–8

Warren, K. *Mineral Resources* (1973)

Wheeler, P. T. 'Ironstone working between Melton Mowbray and Grantham', *East Midland Geographer*, vol 4 (1967), 239–50

Wooldridge, S. W., and Beaver, S. H. 'The working of sand and gravel in Great Britain; a problem in land use', *Geogr J*, vol 115 (1950), 42–57

CHAPTER SIX

Appleton, J. H. 'Some geographical aspects of the modernization of British railways', *Geography*, vol 52 (1967), 357–73

—— *Disused Railways in the Countryside of England and Wales* (1970). Consultant's report for the Countryside Commission

Blake, R. N. E. 'Impact of airfields on the British landscape', *Geogr J*, vol 135 (1969), 508–28

Dower, M. 'Clearance of derelict military buildings', *Journal Town Planning Institute*, vol 45 (1959), 272–5

Hallet, G., and Randall, P. *Maritime Industry and Port Development in South Wales* (Cardiff, 1970)

Hilton, K. J. (ed). *The Lower Swansea Valley Project* (1967)

Miller, T. 'Military airfields and rural planning', *Town Planning Review*, vol 44 (1973), 31–48

Newton, D. E. 'Clearance of concrete from disused airfields', *Journal of the Town Planning Institute*, vol 46 (1960), 169–70

Patmore, J. A. 'The contraction of the network of railway passenger services in England and Wales, 1836–1962', *Transactions and Papers, Institute of British Geographers*, No 38 (1966), 105–18

Pocock, D. C. D. 'Iron and steel at Scunthorpe', *East Midland Geographer*, vol 3 (1963), 124–38

Raistrick, A., and Jennings, B. *A History of Lead Mining in the Pennines* (1965)

Russell, R. *Lost Canals of England and Wales* (Newton Abbot, 1971)

Shankland Cox Partnership (for Southwark BC and the PLA). *A Report on the Future of Surrey Commercial Docks* (1972)

Wallwork, K. L. 'Some problems of subsidence and land use in the Mid-Cheshire industrial area', *Geogr* J, vol 126 (1960), 191–9

—— 'Salt and environmental planning: an historical perspective to a land-use problem', *Proceedings, Fourth Symposium on Salt* (Houston, 1973)

CHAPTER SEVEN

Casson, J., and King, L. A. 'Afforestation of derelict land in Lancashire', *Surveyor*, vol 119 (1960), 1,080–83

Civic Trust. *Reclamation of Derelict Land* (1970)
Clark, S. B. K. 'The task of the government', *Proceedings of the Derelict Land Symposium* (1969), 10–13
Coates, U. A. *Experiments in Grassland Establishment on Colliery Shale, Bickershaw Reservoir Site* (Preston, 1964)
Oxenham, J. R. *Reclaiming Derelict Land* (1966)
—— 'Problems of land reclamation', *Proceedings of the Derelict Land Symposium* (1969), 34–7
Rowbotham, J. 'Landscape reclamation' (review article), *Journal of the Town Planning Institute*, vol 59 (1973), 153
Town Planning Institute. 'The disposal of industrial waste', *Journal of the Town Planning Institute*, vol 36 (1950), 57–61
University of Newcastle upon Tyne. *Landscape Reclamation*, vol I (1971), vol II (1972)
Wise, M. J. 'The Midland reafforesting association 1903–24, and the restoration of derelict land in the Black Country', *Journal of the Institute of Landscape Architects*, No 60 (1962), 13–18

CHAPTER EIGHT

(*a*) *References cited in text*

Appleton, J. H. *Disused Railways in the Countryside of England and Wales* (1970). Consultant's report for the Countryside Commission
Arguile, R. T. 'Reclamation: five industrial sites in the East Midlands', *Journal of the Institution of Municipal Engineers*, vol 98 (1971), 150–6
Barber, E. G. *Win Back the Acres: The Treatment and Cultivation of PFA Surfaces* (1972)
Barr, J. *Derelict Britain* (1969)
Beaver, S. H. 'Land reclamation after surface mineral working', *Journal of the Town Planning Institute*, vol 41 (1955), 146–52
—— 'Land reclamation', *The Chartered Surveyor*, vol 92 (1960), 669–75
—— 'The reclamation of industrial waste land for agricultural and other purposes', in *Problems of Applied Geography* (Warsaw, 1961), 97–112
Cowley, G. *Bedfordshire County Review: Minerals Aspect Report* (Bedford, 1972)
Furness, J. F. 'Tipping in wet pits', *Municipal Journal*, vol 62 (6 August 1954)

Hilton, K. J. (ed). *The Lower Swansea Valley Project* (1967)
James, J. R., Scott, S. F., and Willats, E. C. 'Land use and the changing power industry in England and Wales', *Geogr J*, vol 127 (1961), 286–309
Jones, D. K. C. 'Problems of sand and gravel extraction', *Town and Country Planning*, vol 41 (1973), 368–72
Knabe, W., 'Methods and results of strip mine reclamation in Germany', *Ohio Journal of Science*, vol 64 (1964), 76–105
Land Resource Use in Scotland, HC551 i–v (1972–3)
Langley-Smith, M. 'James Brindley Walk', *Journal of the Institute Landscape Architects*, No 92 (1970), 30–3
McNeill, J. 'The Fife coal industry 1947–67: a study of changing trends and their implications', *Scottish Geographical Magazine*, vol 89 (1973), 163–80
Madin, J. D. H., and partners. *Dawley, Wellington, Oakengates: Consultants Proposals for Development* (1966)
Oxenham, J. R. 'The restoration of derelict land', *Journal of the Town Planning Institute*, vol 38 (1952), 8–12
Parham, E. *Disused Railway Lines in Scotland* (Perth, 1973)
Patmore, J. A. *Land and Leisure* (Newton Abbot, 1970)
Ripley, H. L. 'Analysis of latest experiments in wet-tipping at Egham', *Municipal Journal*, vol 68 (29 January 1960)
Siedlungsverband Ruhrkohlenbezirk. *SVR 1920–1970* (Essen, 1970) (a)
—— *Industriestandort Ruhr* (Essen, 1970) (b)
Thomas, T. M. 'Derelict land in South Wales', *Town Planning Review*, vol 37 (1966), 126–41
Tolley, R. S. 'Telford New Town: conception and reality in West Midlands industrial overspill', *Town Planning Review*, vol 43 (1972), 343–60
Ward, W. J. 'Post publication progress', in Balchin, W. G. V. (ed). *Swansea and its Region* (Swansea, 1971)

(*b*) *Additional references*
Arguile, R. T. 'Some notes on opencast coal mining', *Journal of the Town Planning Institute*, vol 48 (1962), 170–3
Brent-Jones, E. 'Methods and costs of land reclamation', *Quarry Managers' Journal*, vol 55 (1971), 341–51
Bridges, F. M. 'New land for old in the East Midlands', *East Midland Geographer*, vol 4 (1967), 143–53
ECE Symposium on Problems Relating to Environment (United Nations, NY, 1971)

Collinson, G. 'The countryside and the disposal of solid and semi-solid industrial waste and domestic refuse', *Environmental Health*, vol 79 (1971), 191–202

Downing, M. F. 'The reclamation of derelict landscape', *Planning Outlook*, vol 3 (1967), 38–52

Gadgill, R. L. 'Tolerance of heavy metals and the reclamation of industrial waste', *Journal of Applied Ecology*, vol 6 (1969), 247–59

Institution of Civil Engineers. *Civil Engineering Problems of the South Wales Valleys* (1970)

Institution of Civil Engineers. 'Report of the committee on mining and the environment', *Proceedings of the Institution of Civil Engineers*, vol 51 (1972), 637–49

The following references are from *Journal of the Institute of Landscape Architects* (retitled *Landscape Design* in 1972)

Gemmell, R. P. 'Revegetation of toxic sites', No 101 (1973), 28–32

Green B. 'Landscape and the disposal of refuse', No 89 (1970), 26–8

Haywood, S. 'Restoration of landscape: extractive industries', No 91 (1970), 29–30

'Reclamation of ironworks site and slag heaps, Tow Law', No 98 (1972), 24–5

Weddle, A. E. 'The disposal of pulverised fuel ash', No 84 (1968) 30–2

CHAPTER NINE

Beaver, S. H. 'The Black Country', in Myers, J. *Staffordshire*, part 61 of *The Land of Britain* (1945)

—— *Derelict land in the Black Country* (Ministry of Town and Country Planning, 1946 (a)

—— 'The physical background and natural resources', in Abercrombie, Sir L. P., and Jackson, H. *North Staffordshire Plan* (1946) (b)

—— 'The Potteries: a study in the evolution of a cultural landscape', *Transactions and Papers, Institute of British Geographers*, No 34 (1964), 1–31

Burnett, F. T., and Scott, S. F. 'A survey of housing conditions in the urban areas of England and Wales 1960', *Sociological Review*, vol 10 (1962), 35–79

Coates, U. A. *Croal-Irwell Valley Study* (Preston, 1967)

Collins, J. F. N. *Marbury Country Park* (Chester, 1972)

Gulley, J. L. M., and Smith, L. J. 'Aufgelassenes Land, dargestellt am Beispiel der Grafschaft South Lancashire', *Die Erde*, vol 91 (1960), 6–40

Moisley, H. A. 'The industrial and urban development of the North Staffordshire conurbation', *Transactions and Papers, Institute of British Geographers*, No 17 (1951), 151–65

Nicholson, E. M. and Welmacott, R. 'Landscape surgery at Stoke-on-Trent', *New Scientist*, vol 51 (1971), 752–6

North West Economic Planning Council. *Derelict Land in the North West* (Manchester, 1969)

—— *Smoke Control* (Manchester, 1970)

Plant, J. W. et al. *Urban Land Reclamation Project* (1971)

Memoranda submitted by the city of Stoke-on-Trent for the RICS/*The Times* Conservation awards scheme:

'Central Forest Park'

'Westport Water Park'

'Greenways Strategy'

'Clanway Sports Stadium'

'Berry Hill'

Reynolds, J. P. 'The South Lancashire project: a planning exercise', *Town Planning Review*, vol 37 (1966), 102–16

Rimmer, P. J. 'Derelict land in the South Lancashire Coalfield', *Tijdschrift voor Economische en Sociale Geografie*, vol 57 (1966), 160–6

Smith, D. M. 'Identifying the "grey areas", a multivariate approach', *Regional Studies*, vol 2 (1968), 183–93

—— *Industrial Britain: The North West* (Newton Abbot, 1969)

Staffordshire CC. *A Practical Approach to Reclamation* (Stafford, 1969)

—— *Reclamation of Derelict Land* (Stafford, 1970)

'Stoke on Trent reclamation programme', *Journal of the Institute of Landscape Architects*, No 90 (1970), 14–22

Taylor, J. A. 'The Shirdley Hill Sands: a study in changing geographical values', in Steel, R. W., and Lawton, R. (eds). *Liverpool Essays in Geography* (1967), 441–59

Wallwork, K. L. 'Subsidence in the mid-Cheshire industrial area', *Geogr J*, vol 122 (1956), 40–53

—— 'The Mid-Cheshire salt industry', *Geography*, vol 44 (1959), 177–86

—— 'Land-use problems and the evolution of industrial landscapes', *Geography*, vol 45 (1960), 263–75

—— 'Map interpretation and industrial location: the example of alkali manufacture in Lancastria', *Geography*, vol 52 (1967), 166–81

—— 'Salt and environmental planning: an historical perspective to a contemporary land-use problem', *Proceedings, Fourth Symposium on Salt* (Houston, 1973)

CHAPTER TEN

(*a*) *References cited in text*

Bennett, P. M. 'Restoration and reclamation of land by the Opencast Executive of the National Coal Board', *Proceedings of the Derelict Land Symposium* (1969), 38–41

Briggs, R. A. 'Reclamation in County Durham' (British Association for the Advancement of Science, Durham Meeting, 1970). Unpublished

Civic Trust. *Derelict Land* (1964)

Coates, U. A. *Croal-Irwell Valley Study* (Preston, 1967)

Countryside in 1970. *Report on Damage to the Countryside by Industry* (1967)

East Midlands Regional Development Conference. *Report of the Minerals Working Party* (1972)

Ministry of Housing and Local Government. *Derelict Land and Its Reclamation*, Technical Memorandum, No 7 (1956)

—— *New Life for Dead Lands* (1963)

O'Riordan, T. 'Environmental Management', *Progress in Geography*, vol 3 (1971), 173–231

Taylor, M. C. 'Conservation of industrial monuments', *Journal of the Town Planning Institute*, vol 59 (1973), 225–32

Thorburn, A. 'Plunder, pollution or planning?', *Town and Country Planning*, vol 41 (1973), 351–3

Warren, K. 'Yorkshire and Humberside', in Manners, G. et al. *Regional Development in Britain* (1972), Chapter 8

Wibberley, G. P. *Agriculture and Urban Growth* (1960)

(*b*) *Additional references: these works are bibliographies on the reclamation of derelict land:*

Bray, S., and Goodman, G. T. *Ecology of Spoiled and Disturbed Land* (Natural Environment Research Council, 1973)

Department of the Environment (former Ministry of Housing and Local Government). *Derelict Land: A Select List of References, Bibliography No 107, 1930–63*, and *Addenda, 1964–65, 1966–72.* Contains 300 references to dereliction and reclamation in Great Britain

Knabe, W. *Beiträge zur Bibliographie uber Wiederurbarmachung von Bergbauflachen* (Humboldt-Universität, Berlin, 1958)

United States Department of Agriculture. *A Revised Bibliography of Strip-mine Reclamation* (Washington, DC, 1962)

—— *Reference List for Reclamation of Strip-mine Areas* (Washington, DC, 1964)

Vyle, C. J. *Industrial Waste Land – Its Afforestation and Reclamation*, University of Newcastle upon Tyne, Landscape Reclamation Research Project (Newcastle upon Tyne, 1964)

(*c*) *Parliamentary papers*

Parliamentary papers containing references to derelict land are of two main kinds: firstly, annual reports relating to various branches of the mineral and power industries, and secondly, the reports of commissions of enquiry into various aspects of mineral working, land use and pollution. In the first category are the annual reports on mineral production published in various forms since 1854, and the reports of nationalised industries, such as the NCB and the CEGB and its predecessors. The second category includes many reports with incidental references to dereliction and its causes, and others with a more direct relevance. Some of the most important among the last-named group are listed below, in chronological order.

Royal Commission on Mining Subsidence: 1st report (1928), 2nd report (1929)

Report of the Committee on the Restoration of Land Affected by Iron Ore Working (1939)

Report of the Coal Mining Technical Advisory Committee (1945)

Report on the Restoration Problem in the Ironstone Industry in the Midlands (1946)

Report of the Committee on Mining Subsidence (1947)

Report of the Advisory Committee on Sand and Gravel; parts 1–18 (1948–53)

Report of the Technical Committee on Pollution of Water by Tipped Refuse (1961)

Report of the Tribunal Appointed to Inquire into the Disaster at Aberfan (1967)

The Intermediate Areas: Report of a Committee under the Chairmanship of Sir Joseph Hunt (1969)
Report of the Technical Committee on the Disposal of Toxic Solid Wastes (1970)
Royal Commission on Environmental Pollution: 1st Report (1971)

Acknowledgements

PREPARING a work of this kind involves the helpful cooperation of many who, particularly in the public service, have much more pressing claims on their time. I must begin by thanking Professor Richard Lawton both for inviting me to write this book and for his considerable assistance during its preparation.

A very substantial debt of gratitude is owed to those local authority planning offices who put their expert knowledge at my disposal in answering numerous queries and providing data from their files. Fifty authorities were approached, several of them more than once, and only six failed to respond. Particularly sustained support, including the provision of photographs acknowledged elsewhere, has come from the following county planning offices, to whom sincere thanks are due: the counties of Bedfordshire, Cheshire, Cumberland, Derbyshire, Lancashire, Northamptonshire, and Staffordshire; the county boroughs of Manchester, St Helens, and Stoke-on-Trent; and the Monmouthshire Joint Committee. Helpful assistance was also given by the derelict land uints at the Scottish Development Department and the Welsh Office. A particular debt of gratitude is due to Mr S. B. K. Clark of the Department of the Environment, who provided access to the unpublished local authority returns on dereliction, in addition to supplying copies of the annual returns, tracking down out-of-print memoranda and patiently answering many questions on the mechanics of collecting the statistics and defining their terms of reference. In every instance cited above the author alone is responsible for the interpretation of the material employed in this volume.

Assistance was also given by many other institutions and individuals. I would particularly like to thank the staffs of the air photograph libraries at the Department of the Environment (Central air photograph registry), the Ordnance Survey, Fairey Surveys, Meridian Air Maps, Aerofilms and Huntings, and BKS Survey and Technical Services, and of the municipal libraries in Wigan and Wolverhampton. Various sections of the National Coal Board provided useful information, including the Opencast Executive at Harrow, the Statistics Section at Hobart House, the North West Area headquarters at Walkden, and the regional offices of the Minestone Executive at Gateshead. Thanks are also due to the Central Electricity Generating Board and the London Brick Company for information on the disposal of PFA, and to the Sand and Gravel Association of Great Britain. I must also thank Dr G. T. Warwick of the University of Birmingham for drawing my attention to his work on derelict land in the West Midlands, and my colleague Dr J. C. Roberts for drawing my attention to the survey of derelict land in Merthyr Tydfil. Thanks are also due to those colleagues at colleges of education and schools, and their students and pupils, who took part in the questionnaire survey on attitudes to derelict land.

I must also thank members of the technical and secretarial staff in the School of Biological and Environmental Studies at the New University of Ulster, including Miss Shirley Keeley, who drew many of the maps and diagrams from the author's rough drafts. She was assisted by Dominic Loughrey, who with Peter Thompson, also carried out the photographic work. A particular debt of gratitude is owed to Mrs Sally Freeman, who handled a vast amount of preliminary correspondence and also typed the final manuscript, doing both with cheerful efficiency:

without her considerable assistance the preparation of this book would have been immeasurably more difficult.

Although it was not possible to carry out a prolonged programme of fieldwork, assistance from the research funds of the New University of Ulster is acknowledged.

It is a commonplace for authors to thank their families for their forbearance during the preparation of a book. My late wife was a constant source of encouragement, and it is to the memory of her companionship on many excursions to unlovely places that this book is dedicated. My children have long since become accustomed to touring itineraries which never appear in the guide books. I trust, however, that readers will not repeat my daughter's query, voiced after the third circuit of a complex of oil-shale bings, as to why I could not choose something more interesting to write about.

KENNETH L. WALLWORK

Coleraine
September 1973

Index

General note on place-name entries. In order to avoid a plethora of single-entry references to local authority areas in England and Wales only the more important or most frequently mentioned places are cited in the index. References to lesser places are included in the index under the appropriate county. (All names of local authority areas in England and Wales relate to the position on 31 March 1974, ie before the reorganisation of local government.) Additional references to certain areas may also appear under the names of mineral producing localities (eg coalfields).

The photographic plates have not been indexed; reference should be made to the List of Illustrations on pp 7–8.

C

D